About the Author

Norman K. Glendenning was born in Ontario, Canada and took his bachelor's and master's degrees at McMaster University in Hamilton, Ontario. He was awarded a Ph.D. in Theoretical Physics in 1958 by Indiana University. He was promoted to Fellow of the American Physical Society in 1968 and won the Prestigious Alexander von Humboldt Prize in 1994. He has spent his entire academic career at the University of California's Lawrence Berkeley National Laboratory. Besides publishing more than 200 research papers in scientific journals, Dr Glendenning has written two previous books concerning his research: *Direct Nuclear Reactions* and *Compact Stars*.

Ptolemy of Alexandria charting the constellations. *Reproduced with permission from Bibliotheque National de France.*

Norman K. Glendenning

Lawrence Berkeley National Laboratory, USA

AFTER THE
BEGINNING

A Cosmic Journey through Space and Time

World Scientific

Imperial College Press

Published by

Imperial College Press
57 Shelton Street
Covent Garden
London WC2H 9HE

and

World Scientific Publishing Co. Pte. Ltd.
5 Toh Tuck Link, Singapore 596224
USA office: 27 Warren Street, Suite 401–402, Hackensack, NJ 07601
UK office: 57 Shelton Street, Covent Garden, London WC2H 9HE

British Library Cataloguing-in-Publication Data
A catalogue record for this book is available from the British Library.

Cover photo: Deep view of space from the Hubble Space Telescope. In this single view we see exactly what Laplace divined (see page 3). Remarkably, he lived centuries too early to see—as we can see now—the disk shapes that galaxies develop as they evolve under the influence of gravity and angular momentum conservation. *Photo: Hickson Compact Group 87 (HCG 87); courtesy of NASA.*

AFTER THE BEGINNING: A COSMIC JOURNEY THROUGH SPACE AND TIME

ISBN 1-86094-447-7
ISBN 1-86094-448-5 (pbk)

Printed in Singapore by Mainland Press

To my dear children:

Nathan, Elke, Alan

Contents

Preface

I know not what I appear to the world, but to myself I seem to have been only like a boy playing on the sea-shore, and diverting myself in now and then finding a smoother pebble or a prettier shell, whilst the great ocean of truth lay all undiscovered before me.

— Isaac Newton, *Memoirs*

The heavens are a wonder to us all. So it has been throughout the ages. When we look at the night sky we see what seems a timeless panorama of stars. The slow procession of the planets and an occasional shooting star suggest a sedate motion in an otherwise eternal and unchanging universe.

It isn't so. It has long been known that stars are constantly being born in great gaseous clouds, that they develop in complexity over millions of years and then eventually die. Indeed, in the early years of the last century, the famed British astronomer Sir Arthur Eddington described the Sun as a great furnace that had enough fuel to burn for 12 billion years before it would fade away. His estimate is sound, according to all the developments in the understanding of stellar processes since then. This much is commonly known, if not the details.

What is less well known, which was only recently confirmed by the many different ways in which the heavens can be viewed by modern instruments, many of them based on satellites or carried aloft by balloons, is that from the earliest moments the universe has been a cauldron of fiery activity.

At the beginning, the fire was so intense that nothing in the universe now resembles what it was made of then. The entire *part* of the universe that astronomers can possibly see — limited as it is, not by their instruments alone, but by the distance that light can travel since the beginning — was contained in a very small space. From such a beginning, how did the universe evolve to make stars and the elements out of which planets could be made and from which life could emerge? This is the story I wish to tell. Still more, I include brief stories of some of the men and women who have revealed the cosmos to us. As David Knight wrote in his preface to Rupert Hall's *Isaac Newton*, "Science is a fully human activity; the personalities of those who practice it are important in its progress and often interesting to us. Looking at the lives of scientists is a way of bringing science to life."

I write especially with the layman in mind; for the more technically inclined I have placed interesting derivations and calculations in boxes at the end of chapters. In this way I think I have written a story of our universe that will be satisfying to the lay reader as well as to the scientist who would like to become more familiar with a subject — the cosmos — that, beginning in childhood, fills us all with wonder.

The universe we live in is as beautiful as it is awesome — more so to me, having with great pleasure learned enough to write these pages. I hope they will give pleasure to the reader. These are wonderful times in the history of science *on this planet* to be a cosmologist.

Norman K. Glendenning
Lawrence Berkeley National Laboratory
November 2001

Acknowledgments

Many thanks to Laura Glen Louis for encouraging me to write this book for the children and especially for the proofing and the many valuable suggestions, and to Jean-Pierre Luminet, whose many publications have been an inspiration.

I also appreciate the help in various forms from Jorg Hufner (Heidelberg), Saul Perlmutter (Berkeley), Sanjay Reddy (Los Alamos), Martin Redlich (Berkeley) and Wladeslaw Swiatecki (Berkeley).

List of Figures

Hierarchy of Cosmic Structures

Object	Size	Mass
Sun	7×10^5 km	2×10^{33} g $\equiv M_\odot$
Galaxy	45×10^{16} km	$10^{11} M_\odot$
Galaxy cluster	15×10^{19} km	$10^{13} M_\odot$
Supercluster	150×10^{19} km	$10^{15} M_\odot$
Universe	$10\,000 \times 10^{19} = 10^{23}$ km	$10^{21} M_\odot$

Timeline of Particle Appearances

Cosmic Content	Temperature (K)	Time (seconds)
Quantum foam	4×10^{32}	10^{-43}
Horizon < nucleon size	6×10^{21}	3×10^{-24}
$\gamma, \nu\bar{\nu}, e\bar{e}, q\bar{q},$ Z^0, W	10^{15}	10^{-11}
$\gamma, \nu\bar{\nu}, e\bar{e}, q\bar{q},$	10^{14}	10^{-8}
$\gamma, \nu\bar{\nu}, e\bar{e},$ p, n, hyperons	10^{12}	10^{-5}
$\gamma, \nu\bar{\nu}, e\bar{e},$ p, n	10^{10}	1
$\gamma, \nu\bar{\nu},$ H, D, He, Li (ions)	10^9	100
Atoms of the above (recombination)	3000	$300\,000$ yr
$\gamma, \nu\bar{\nu},$ H, D, He, Li $\cdots$ Pb	180	10^8 yr
Galaxies, stars, planets, life	2.73	15×10^9 yr

1 Island Universes

After dinner, the weather being warm we went into the garden and drank tea, under the shade of some apple trees. Amongst other discourse, he told me...the apple draws the earth, as well as the earth draws the apple...there is a power, like that we call gravity, which extends its self thro' the universe.

— from a conversation of Isaac Newton with and reported by
W. Stukeley, in *Memoirs* (1936)

1.1 The First Move

When I was a child, I read that God created the heavens and the earth, the dry land and the sea.... And on the first day, He separated day from night. He commanded that the waters bring forth abundantly the moving creatures...the cattle, and every creeping thing, each after its own kind.... On the sixth day He made man, male and female, in His own image. And God saw that it was good.

When I grew to be a boy I asked myself: "Who made God?" I could find no answer. Some spoke of an everlasting God, but that was as hard to imagine as a God who had a beginning. I turned to the philosophers and discovered that the question has been pondered through the ages. From what I have since gathered, every culture has puzzled over the beginning, and many myths have been told.

Perhaps the beginning can never be fathomed. But what happened *after the beginning* is the subject of this book. From the evidence that has been gathered here on the Earth and in the heavens by giant telescopes and sensitive apparatus carried aloft in balloons and satellites, cosmologists have been able to piece together the story of the universe and how — in one brilliant moment about 15 billion years ago — time was born in the instant of creation of an immensely hot and dense universe. From such a fiery beginning, the universe began its rapid expansion *everywhere*, creating space

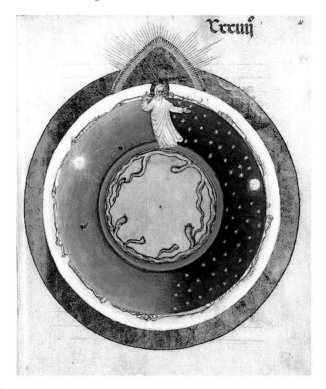

Fig. 1.1. And on the first day He separated the day from the night.... *Credit: A miniature belonging to the artistic period of Lombard in the 14th century, depicting the third day of creation. It was included in an exposition, "Figures du Ciel", at the site Francois Mitterrand, Grande Galerie, Paris (9 Octobre 1998–10 Janvier 1999). Other parts of the exposition can be viewed on the WWW at http://expositions.bnf.fr/ciel/mythes/index.htm*

on its outward journey where there was no space, and time where there was no time.

Through the known laws of nature, scientists are able to chart the course of cosmic history in quite some *verifiable* detail. Now, we have a basic understanding of how, in the course of these billions of years, the hot and formless primordial clouds of matter and radiation slowly cooled and began to collapse under the universal force of gravity to form galaxies of stars and billions of planets. It was truly a miracle. And it is that miracle which I wish to tell of.

But how that greater miracle — life itself — arose on the Earth, no the one seems to know. Yet, if the miracle happened on this planet, then it surely also happened on others far away. For, in our galaxy alone, an estimated *hundred billion* planets circle other suns.[1]

Even as children, we try to penetrate the mysteries. *How, when, and why?* Was there a time before matter, or did time and matter come into existence together? *How* did life arise from the inanimate world, eventually to form the human minds that contemplate and seek the answers? These are deep mysteries that perplex us.

Fig. 1.2. M17, also known as the Omega or Swan Nebula, is a star-forming region located about 5500 light years away in the constellation Sagittarius. *Credit: NASA, ESA, and J. Hester (ASU).*

Laplace (1749–1827) also pondered these mysteries and provided the direction of an answer to one of the three questions. There *was* a beginning. He understood that at one time, long ago, a great cloud of gas must have contracted under the influence of its own gravity to form our Sun, our Earth, and all the other planets and comets of the solar system. As these clouds contract and fragment, their angular velocities must increase so as to conserve their angular momentum, just as an ice-skater who goes into a whirl with arms outstretched and then draws them in, whirls faster. The whirling motion of the gravitating nebulous objects flattened them into disks and at the same time prevented their total collapse.

We see this almost universal tendency of rotating bodies that are held together by gravity. They eventually form disks with central bulges, such

---

[1] "Detection of extrasolar giant planets", G.W. Marcy and R.P. Butler, *Annual Review of Astronomy & Astrophysics*, Vol. 36 (1998), p. 56.

Fig. 1.3. An edge-on view of a warped spiral galaxy (ESO 510-G13) that is otherwise similar to our own. *Credit: NASA and STScI/AURA; J.C. Howk (Johns Hopkins University) and B.D. Savage (University of Wisconsin–Madison).*

as the shape of the solar system and of the entire galaxy. Indeed, as the astronomer peers deep into space — and therefore back in time — he sees that all galaxies rotate and are disks (Figure 1.3). *But* clusters of galaxies, though they too rotate, are very irregular in shape — not at all like disks.[2] There has not been sufficient time for them to have flattened into disks. Therefore, the universe has not lived forever, but had a *beginning*.

Cosmology is the subject of this book — the events that shaped the universe from its early moments to the present time as well as its possible futures. There are numerous diversions in the telling of this story, because there are so many interesting men and women who have provided pieces of the evidence, and they are interesting too.

1.2   Looking Back

> And the same year I began to think of gravity extending to the orb of the Moon. . . .
>
> — Isaac Newton, *Waste Book* (1666)

Until recently, most scientists regarded cosmology as having little foundation in fact, a somewhat disreputable subject that serious people might

---

[2]J. Huchra and M. Geller, 1998.

Fig. 1.4. The evolution of the universe has fascinated us from time immemorial. Virtually every society has its creation myth. Shown here is the first of the modern cosmologists — the young P.J.E. Peebles, from the plains of Manitoba to the halls of Princeton, there to work with Robert Dicke. *With kind permission of P.J.E.P.*

well ignore. Nevertheless, some very serious people did pursue what evidence there was about the world's beginning and were finally rewarded (Figure 1.4). The last several decades have been unparalleled in the growth of our knowledge of the physical and biological worlds and in the creation of new and unforeseen technologies. No one could have dreamed that in less than the span of a single lifetime we would learn that from a primeval fireball the universe would enter an era of *accelerating* expansion. Let us glance in passing at the very early cosmologists and naturalists who led the way.

An advanced Babylonian culture emerged in the fertile lands called Mesopotamia, between the Tigris and Euphrates rivers in southwest Asia.[3] By 3000 B.C. these people had developed a system of writing, which facilitated their explorations in science and mathematics. They had begun to delve into the oldest of sciences — astronomy — as early as 4000 B.C. These early scholars — who, as in other ancient cultures, were priests — recorded their findings on thousands of clay tablets. The tablets bore observations and calculations of the motions of the planets with which they

---

[3]Presently known as Iraq.

could predict eclipses and lunar periods. Their best work was done in the period following the destruction of the Assyrian capital city of Nineveh in the seventh century B.C., and it continued until about the birth of Christ. However, because they lacked knowledge of geometry and trigonometry, they did not advance as far as the Greeks in their astronomical studies.

The early Greek philosopher Pythagoras (569–475 B.C.), born on the island of Samos, is believed to be the first pure mathematician. He was interested in the principles of mathematics, the concept of numbers, and, very importantly, the abstract idea of a proof. Later philosophers, Plato (428–347 B.C.), and his student in the Athenian academy, Aristotle (384–322 B.C.), tried to understand the world they saw in terms of natural *causes*, a concept that was novel at the time and later lost until Galileo and Newton reintroduced it 2000 years later. Aristotle realized that the motions of the Sun, Moon, and planets across the sky and the shadow of the Earth on the Moon were evidence that the world is round but mistakenly believed that the heavenly bodies circle the Earth.

Claudius Ptolemy (85–165 A.D.) of Alexandria, renowned astronomer, mathematician, and geographer, proposed that the Sun and other planets were fixed on giant celestial spheres that rotated about the Earth in the order shown in Figure 1.5. This system became known as the Ptolemaic system. It was so successful at predicting the motion of the planets that it became somewhat of an obstacle to further progress, for it seemed to many to provide a satisfactory account of the heavens.

Over the course of the 20 centuries following Pythagoras, the only person known to have questioned these Earth-centered cosmologies was an early Greek philosopher named Aristarchus (310–230 B.C.). Born on the island of Samos, Aristarchus taught that the Sun was the center of the universe and that the planets, including the Earth, revolved about the Sun. However, other philosophers, steeped in the writings of Pythagoras, Plato, and Aristotle, persisted in the tradition of placing the Earth at the center.

But all this was swept away and lost for generations when a powerful Macedonian dynasty wrested control of the Greek city states. To King Philip and his wild pagan princess, Olympias of Epirus, Alexander (356–323 B.C.) was born. The child's education was strict and rigorous, first under the tutorship of his cunning and ferocious mother and a loyal old soldier kin of hers. When the lad entered his teens, Philip brought Aristotle from Athens to tutor his son and a lasting bond was forged. At the age of 16, Philip put Alexander in command of the left wing of his forces, where the Athenians had drawn up before them a great host to protect their city. A strategic retreat by Philip on the right opened a weakness in the

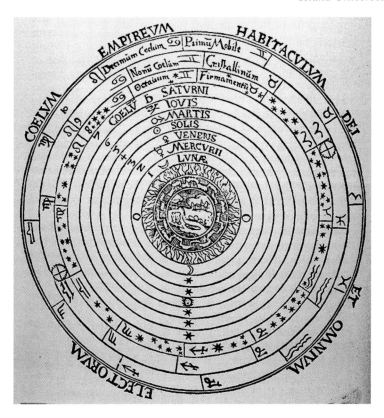

Fig. 1.5. The Earth-centered heavenly arrangement of the planets, Sun, and Moon in Ptolemy's astronomy. *Credit: From Peter Apian,* Cosmographia *(1524).*

Athenian center into which Alexander at the head of his cavalry plunged with devastating effect. One by one the free Greek city states came under the protectorship of Philip.

Following Philip's assassination, Alexander, later known as "the Great", quickly reaffirmed the authority of the Macedonians. The great tide of Alexander's conquests that soon swept over Asia and India often obscures what is important to our story of cosmology — an end to the independence of the Greek cities and to the spirit of free inquiry of their citizens. That freedom lost, the riches of the Babylonian and Greek heritage — in cosmology, government, philosophy, ethics, and literature — survived only in manuscripts. But the flickering flame, rekindled for several centuries, was lost again when the barbarians swept into Europe in the fourth century and the Dark Ages fell over the land. Meanwhile, the Roman Church revived a strict interpretation of the scriptures, insisting that God made a flat Earth with a heavenly canopy of fixed stars.

Greek learning began to seep into Europe in the 12th century through contacts with Muslim Spain, where Arabs had built on the Greek legacy and Indian mathematics. This trend accelerated in the 13th century with the growth of wealth through banking and foreign trade in some Italian city states, especially Florence, Venice, Milan, and Rome. Wealthy citizens like the Medici sent emissaries abroad to collect early Greek manuscripts, both in the original and in translations by Arabs, as well as later Roman writings. Scholars, artists, engineers, and bankers throughout Europe were amazed and inspired by the accomplishments in the arts, sciences, logic, and government that had preceded them by 2000 years. In Florence alone, Fra Angelico, Boticelli, Brunelleschi, Donatello, Leonardo, and Michelangelo thrived. A new age was born — the Renaissance.

Meanwhile, in Poland, Copernicus (1473–1543) found a simpler, more elegant account of the motion of the Sun and planets than was provided by the cosmologies of the early Greeks and their successors. Copernicus achieved his goal of simplification by allowing the Earth and other planets to move in their separate circular orbits around the Sun. Meanwhile Tycho Brahe (1546–1601) in Denmark built and calibrated accurate instruments for his nightly observations. Tycho observed the new star in Cassiopeia (the "queen") — a supernova, and the most recent star to explode in our own galaxy. He recorded the discovery in a short paper. He had his own printing press, which he used to record his observations of the motions of planets and their moons. Tycho hired Johannes Kepler, born in Württemberg, a few kilometers to the north of Heidelberg, to be his assistant.

When Tycho died, Johannes Kepler (1571–1630) inherited his post as Imperial Mathematician in Prague. Using the data that Tycho had collected through his observations at the telescope, Kepler discovered that the planets move not on perfect circles but on elliptical orbits with the Sun at one focus. He found that their speed increased as they approached the Sun, and slowed as they receded, and he deduced that their motion swept out equal areas in equal times. We know these achievements as Kepler's laws of planetary motion. Indeed, based on these laws and on Tycho's observations he calculated tables (the Rudolphine tables) of the orbital motion of the planets that turned out to remain accurate over decades.

In 1609 Kepler published *Astronomia Nova*, announcing his discovery of the first two laws of planetary motion. And what is just as important about this work, "it is the first published account wherein a scientist documents

Fig. 1.6. *Left:* Johannes Kepler (1571–1630), discoverer of the laws of planetary motion, which Newton later used to confirm his theory of gravitation. *Credit: Aus dem Museum der Sternwarte Kremsmunster in Prague. Right:* Portrait of Galileo (1564–1642), thought to be by Tintoretto. *Credit: London, National Maritime Museum.*

how he has coped with the multitude of imperfect data to forge a theory of surpassing accuracy".[4] Kepler was the first astronomer to think about the physical interpretation of the celestial motions instead of merely observing and recording what he saw.[5] His accomplishments were profoundly important in establishing the Sun as the center of the planetary system, thus helping to confirm the Copernican world view that Galileo (1564–1642) later used to buttress his own discoveries.

A Dutch spectacle maker applied in 1608 for a patent on a revolutionary tool — the telescope. But within several weeks, two others also made claims, and the patent officials concluded that the device was useful but too easily copied to warrant a patent. The news of the telescope spread quickly over Europe through diplomatic pouches. Rumors soon reached Venice, where

---

[4] O. Gingerich in the foreword to *Johannes Kepler New Astronomy*, translated by W. Donahue (Cambridge University Press, 1992); O Gingerich, *The Eye of Heaven: Ptolemy, Copernicus, Kepler* (1993).

[5] A. Rupert Hall, *Isaac Newton: Adventurer in Thought* (Cambridge University Press, 1992).

a priest sought confirmation by a letter written to a former student of Galileo Galilei[6] then residing in Paris. The student confirmed the availability of spyglasses there. The priest showed the reply to his friend Galileo. In his marvelous book called *Siderius Nuncius*,[7] relating many astronomical observations, Galileo reported that on the very night following his return to Padua from Venice, he had constructed his own instrument.

The first telescopes made in Holland and Paris used lenses made by opticians for eyeglasses; they had a magnification of only three. Galileo set about constructing more powerful instruments. After some experimentation, he determined that he needed a weaker convex lens in combination with a stronger concave lens, neither of which was available in optical shops. He therefore set about teaching himself the fine art of grinding lenses. Soon he had a telescope in hand that could magnify nine times. Only a year following the patent application in Holland, Galileo turned his own, more powerful telescopes toward the heavens. He realized that the telescope would revolutionize astronomy and cosmology, and he set about this task with gusto.

After perfecting an instrument that would magnify 20 times, he began numerous systematic observations which he recorded in drawings, notes, and published books.[8] One of them, *Discourses on Two New Sciences*, was written in 1638 after his trial before the Inquisition. The manuscript was smuggled out of Italy to Leiden for publication.

Galileo perceived through his telescope that the Moon was not the perfectly smooth "celestial sphere" as all heavenly bodies were thought to be, but rather it was rough: its surface was broken by mountains and valleys, as revealed by the shadows that the Sun cast (Figure 1.2). Important among other discoveries that contradicted Aristotelian and Church belief in the central role of the Earth were the phases of Venus. Galileo realized that the phases resulted from Venus' orbit about the Sun, which illuminated

---

[6]Galileo was born in Pisa in February 1564, died in his home at Arcetri in the hills above Florence in January 1642, and is buried in Santa Croce. He is the last scientist to be known to the world by his first name.

[7]*Siderius Nuncius* means "Starry Messenger"; its actual cover of 1610 is shown in Figure 1.7.

[8]Galileo Galilei (1564–1642), *Siderius Nuncius*, published in Venice, May 1610; translated by Albert Van Helden (University of Chicago Press, 1989). *Dialogue Concerning the Two Chief World Systems*, published in Florence, February 1632, translated by S. Drake, 2nd edition (University of California Press, 1967). *Discourses on Two New Sciences*, published in Leiden, 1638; translated by S. Drake (University of Wisconsin Press, 1974).

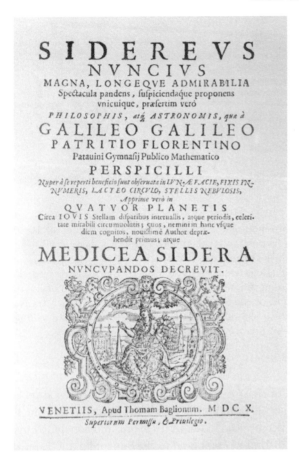

Fig. 1.7. Cover of Galileo's *Siderius Nuncius* ("Starry Message or Messenger"), in which he published many of his astronomical discoveries made with telescopes of his own construction. Prominently announced on the cover is his position as philosopher and mathematician, his station in society as a Florentine patrician, and importantly, his discovery of four of the moons of Jupiter, which he names the Medician stars, in honor of his hoped-for (and soon-to-be) patron. Note the place and date on the bottom line (Venice, 1610). The book's appearance in print followed his first acquaintance with telescopes by less than two years. *With kind permission of Instituto e Museo di Storia della Scienza, Firenze.*

first one face, and then the other. He observed the four Medicean moons of Jupiter and their individual orbits around that planet, which provided further evidence that planets and moons circled other bodies in the heavens, while only one moon circled ours.

 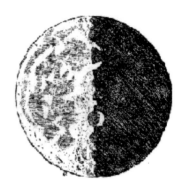

Fig. 1.8. Reproductions of some of Galileo's drawings of the Moon showing that its surface is not the ideal smooth celestial sphere of all heavenly bodies as held by Church doctrine, passed down from Aristotle, but had mountains and valleys whose shadows were easily identified. Galileo perceived that the lighting on opposite faces at different phases proved that the Moon orbits the Earth. From *Sidereus Nuncius, or the Sidereal Messenger*, Galileo Galilei; translated by Albert Van Helden (University of Chicago Press, 1989). *Reproduced with permission from Wellesley College Library, Special Collections.*

According to Aristotelian and Church teaching, the stars lay fixed on an enclosing membrane around the Earth not so far beyond the Sun. However, Galileo could perceive no diameter to a star and concluded instead that the stars are very distant as compared to the Sun and planets, for they remained but points of light in his telescope. All of these and his numerous other discoveries convinced him that moons orbited planets and planets orbited the Sun; he became a champion of Copernicus' Sun-centered cosmology.

What it was that actually guided the motions of the heavenly bodies had no scientific foundation until the time of Isaac Newton (1643–1726). He hypothesized that "...all matter attracts all other matter with a force proportional to the product of their masses and inversely proportional to the square of the distance between them". The falling apple and the Moon are both impelled in their motion by the force of the Earth's gravity. The apple falls down because it hangs motionless above the Earth as together they rotate. The Moon also falls toward the Earth, but because of its tangential velocity it does not fall down but falls around (see Box 1). Newton's theory was the first to account for the motion of bodies by forces that could be measured or predicted, given the appropriate data. It marked the birth of modern scientific cosmology.

Fig. 1.9. A mask of Sir Isaac Newton taken after his death. "Plato is my friend, Aristotle is my friend, but my best friend is truth." (From his notebook in Latin which he titled "Certain Philosophical Questions".) *By kind permission of the Provost and Scholars of King's College, Cambridge.*

Sir Isaac Newton, together with Albert Einstein (1879–1955), was an intellectual giant. Newton was born a year after Galileo died, in Lincolnshire into a family of fair means, though his father died before he was born. His mother remarried, and young Isaac was raised by his grandmother. None of his father's estate passed to him until his mother's death. Meanwhile he was admitted to Trinity College, Cambridge when he was 18. The journey from his home down the Great North Road in 1661 took three days. From her considerable wealth, his mother, an undemonstrative woman, provided a meager allowance to the young student. Renowned as Cambridge is in our day, in Newton's it was a place either of idleness or of intellectual liberty. For Newton, it was clearly the latter. From his allowance he bought many books and borrowed others from his professor, Isaac Barrow, making extensive memoranda, including original ideas and calculations, in his voluminous notebooks. He read, in Latin or Greek, works by Descartes, Huygens, and many other scholars, Henry Moore, Galileo, and the long didactic poem *On the Nature of Things*, by the Greek atomist Lucretius (about 96–55 BC).[9] It is to Lucretius that we owe the word "atom". Newton, ever brimming with ideas, sometimes extended his notes from what he had read, to record original research, both experimental and mathematical.

Although appointed at the age of 27 to the Lucasian professorship at Cambridge,[10] his first scientific publication was severely (and inappropriately) criticized by a prominent member of the Royal Society, for whom he nourished thereafter a lifelong animosity.[11] Perhaps as a result of that episode, he appears to have experienced a deep anguish throughout his life: this desire for fame and recognition could come only with the publication of his work, but his desire conflicted with a fear of criticism and plagiarism if his work became known. In any case, his productive life as a scientist and mathematician ended when he left Cambridge at the age of 53 to take up a civil service position in London. The post was an important and powerful one. Newton was elected president of the Royal Society in 1703, and retained this honor until his death at the age of 84. He was knighted by Queen Anne in 1705.

As Warden and later Master of the Mint, the integrity of the nation's coinage, and with it that of the banks on which all foreign trade relied, was in Newton's hands. The nation's trading status was weakened to the extent that counterfeiters produced coins with a deficit of gold content. At Newton's instigation, parliament had elevated counterfeiting to an act of treason. Thereupon, Newton turned his magnificent intellect and unrelenting determination to ending the practice. He sought and received compensation from his superiors for disguises in which he frequented the low-life taverns of London, making notes as he eavesdropped on careless and boastful revelers. Eventually he had accumulated sufficient hearsay evidence against a few unfortunates, who, when confronted with so much information about themselves, became pliant witnesses against the kingpins.

During Newton's years at Cambridge, the quality of University appointments was severely threatened by King James II, who had earlier converted to Roman Catholicism. James became fearful of his Protestant subjects and began to appoint only Roman Catholics as officers in his army, as judges, and as professors at Oxford and Cambridge, without regard to their qualifications. Newton strongly opposed what he saw as an attack on the

---

[9]Lucretius' ancient and prescient work had been rediscovered in Newton's time.

[10]The chair is now held by Stephen Hawking.

[11]Biographical notes are excerpted from A. Rupert Hall, *Isaac Newton: Adventurer in Thought* (Cambridge University Press, 1992).

freedom of the University and urged the Vice-Chancellor to "Be courageous and steady to the Laws and you cannot fail". The King dismissed the Vice-Chancellor, but Newton continued to argue vigorously for the defense of the University against the King. In the meantime, leaders of the political opposition invited William of Orange to bring an army from Holland to defeat James. James fled to France, and Cambridge elected Newton to the Parliament in 1689. Parliament offered the crown to William and Mary.

Newton's discoveries and accomplishments included gravity, the laws of mechanical motion and celestial mechanics, differential and integral calculus, and optics. From the time of Aristotle, white light was believed to be a single entity — a color like the others. Newton showed that white light was composed of all the colors of the spectrum by passing a ray of sunlight through a glass prism. He argued that light was corpuscular in nature, contrary to the prevalent view that it was wavelike, although he also employed properties of rays to understand some of his experiments. We now know that light has properties that are characteristic of *both* particles and waves. He found that refracting telescopes of his day were limited in their resolution because different colors were refracted by slightly different angles (chromatic aberration) and so could not make a precisely sharp image. In response to this limitation Newton perfected the reflecting mirror telescope, which is now universally employed by astronomers when observing objects in the universe with an optical telescope.

Eleven years after Newton's death, the man who was to become the most renowned astronomer of his time, William Herschel (1738–1822), was born into a family of modest means in Hanover. He immigrated to England in 1759 when he was only 21, to escape the seven years' war between Prussia and Hanover with France after experiencing the battle at Hastenbeck, where commanders on each side thought that they had lost the battle.

Herschel took up a successful career in music, eventually becoming a teacher and organist at the Octagon Chapel in Bath. Inspired by the atmosphere of inquiry that was fostered by his father in his early family life, he became interested in astronomy, at first renting telescopes and then building his own. With a grant from King George III, he built the best and most powerful instrument the world had seen, a 48-inch reflecting telescope based on Newton's design. Herschel became a full-time astronomer when George III knighted him and appointed him Astronomer Royal with a pension.

How thrilling it must have been for Herschel and his sister to discover other worlds far beyond our own Milky Way galaxy! While perched on a ladder near the end of his powerful but unwieldy instrument, Herschel would call out a description of the objects that interested him to his sister,

Caroline, on the ground below. As the night progressed and the Earth turned, a thin slice of sky from east to west came into the view of the fixed position of his telescope. Each day he changed the angle of the telescope slightly, and in this way brother and sister recorded the description and positions in the sky of nearly 5000 objects. With further additions of faint patches of light called nebulae, many of which were galaxies, the revised catalogue of stars and nebulae became known as the General Catalogue (1864). It was later extended and is now known as the New General Catalogue or simply as NGC. Galaxies and other nebulous objects that were discovered then and later are still known by their number in the catalogue, such as NGC 3132, the exploding star shown in Figure 1.2, the center of which is becoming a white dwarf.[12] Such will be the fate of our Sun together with the conflagration of the entire solar system in about seven billion years.

Caroline Herschel (b. Hanover, 1750–1848) had joined her brother, William, in England while he was still a musical conductor. Their father, a musician, though with no formal education, was himself interested also in philosophy and astronomy and encouraged his children to explore nature. Caroline records that her father took her "...on a clear frosty night into the street, to make me acquainted with several of the beautiful constellations, after we had been gazing at a comet which was then visible". In England, she trained as a singer but slowly returned to her early acquaintance with comets. She studied mathematics under William, especially spherical trigonometry; she employed this to reduce the data that she and her brother gathered during their nights at the telescope.

Caroline Herschel discovered her first comet shortly after William gave her a telescope of her own. There followed quickly a number of other discoveries, including M110, one of the satellite galaxies of the Andromeda (Figure 1.12), as well as NGC253, a galaxy in the direction of the constellation Sculptor. It was not until later that these nebulous objects were actually perceived as galaxies containing individual stars.

In the meantime Caroline and Sir William continued their compilation of nebulae for the New General Catalogue. With the award by George III in 1787 of an annual salary to continue as William's assistant, Caroline became the first woman officially recognized with a scientific position. After her brother's death, she continued her research for many years, becoming a renowned scientist in her own right. Caroline Herschel was awarded

---

[12] A French astronomer, Messier, interested primarily in comets, had earlier found a number of nebulae which he recorded in his catalogue. Objects in that catalogue are known by the letter M followed by a sequence number, such as M31 shown in Figure 1.12.

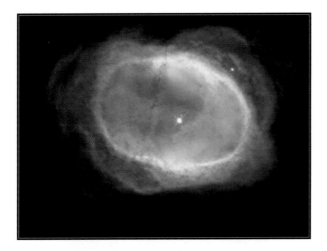

Fig. 1.10. The Southern Ring (NGC3132), a *planetary nebula*, is the remains of a star of about our Sun's mass, which, near the end of its life, expands into a very much larger and bloated red giant. A similar fate awaits our Sun in about seven billion years. The red giant expels hot gasses from its surface at velocities of tens of kilometers a second, and these hot ionized gasses are what is seen as the planetary nebula. It will dissipate and disappear into interstellar space after some 10 000 years. The ember of the dying star in the center will cool for billions of years as a white dwarf star with a radius of about 1000 km. After that the nebula will simply fade from view, and much later, on the order of the present age of the universe, the ember may crystallize into a diamond with some impurities such as magnesium and an atmosphere of hydrogen, helium, and oxygen. *Image credit: NASA and The Hubble Heritage Team (STScI/AURA). Acknowledgment: R. Sahai (Jet Propulsion Lab).*

numerous honors, including the Gold Medal of the Royal Astronomical Society and the Gold Medal for Science by the King of Prussia. On her 97th birthday she was an honored guest of the crown prince and princess of Prussia.[13] She lived an active life to the end.

Just how distant and numerous the stars are, was not imagined until William Herschel speculated in the late 1700s that some of the hazy patches of light that could be seen among the stars through his telescope were actually distant "island universes". He proposed that the Milky Way was also such an island of stars in a vast empty space. It wasn't until the early

---

[13] A.M. Clerke, "Caroline Lucretia Herschel", *Dictionary of National Biography XXVI* (London, 1891), pp. 260–3.

Fig. 1.11. Caroline Herschel (1750–1848), singer and amateur astronomer, the first woman to be salaried as an astronomer (by George III of England), winner of the Gold Medal of the Royal Astronomical Society, and of the Gold Medal for Science by the King of Prussia, discoverer of many galaxies and nebulae. *Engraving by Joseph Brown, 1847. Credit: National Portrait Gallery, London.*

1900s that astronomers were actually able to observe the individual stars in nearby spiral galaxies using larger telescopes.

With the discovery that our Sun is but a minor star among many billions in the Milky Way, and that our galaxy is but one among many, now known to number also in the billions, our perspective on our place in the universe has changed forever. We can no longer entertain the notion that we are the center of it all. We can be humbled by this fact and at the same time marvel that the forces of nature, acting over billions of years, have made such a beautiful and perhaps boundless universe, and that our kind is here to live in it and to try to comprehend it for, perhaps, a few more thousand years. . . .

1.3   Age and Size of the Visible Universe

> If I have been able to see further, it was only because I stood on the shoulders of giants.
>
> — Isaac Newton, from a letter to Robert Hooke

Fig. 1.12. The Andromeda galaxy (M31) was first recorded by the Persian astronomer Al-Sufi (903–986 A.D.), who lived in the court of Emire Adud ad-Daula. He described and depicted it in his *Book of Fixed Stars* (964 A.D.) and called it the "little cloud". The galaxy is composed of about 400 billion stars. It lies relatively close to our own galaxy, the Milky Way, and is rather similar, having a central bulge and flat disk in the form of spiral arms. "Relatively close" in this context means 2 900 000 light years (or, equivalently, a thousand billion kilometers). Two smaller elliptical galaxies, M32 and M110 (the two bright spots outside the main galaxy), are in orbit about Andromeda. The myriad foreground stars are in our own galaxy. *Credit: George Greaney obtained this color image of spiral galaxy M31, the Great Galaxy in Andromeda, together with its smaller elliptical satellite galaxies.*

It is marvelous how much astronomers have learned about our universe. So many have contributed. Much of what has been discovered seems puzzling at first or is not understood at all. But, like a jigsaw puzzle, the separate pieces, often meaningless in themselves, begin to make a design. And when many of the pieces fit, one draws confidence from the perception of a larger design and it becomes easier to fit the other pieces into place. So it is with our understanding of the universe. It is not complete. But a grand picture is emerging, if not all the details. Let us look at some of the pieces first.

Light does not move instantaneously from one place to another. It travels at the constant speed of 300 000 kilometers per second. That is fast, but not instantaneous. Because the speed of light is finite, when the astronomer peers at galaxies spread out in space, he does not see them now in the very instant of his gaze, but rather, *he sees them as they were at some earlier*

Fig. 1.13. The Orion Nebula (M42) was first seen as a cluster of stars by John Herschel, son of Sir William Herschel. John used a telescope built by father and son, during his exploration of the southern sky from Cape Town. Orion is quite close to our Milky Way galaxy — about 17 Milky Way diameters in distance — and can be seen with the naked eye. The inner regions glow in the red light of excited hydrogen, which, with some green emission from oxygen, produces a yellowish color in the center of the nebula. The energy for this spectacular display comes from the small cluster of stars in the brightest part of the nebula. *Copyright of Anglo-Australian Observatory and photograph by David Malin.*

time, each according to its distance. The deeper he looks, the further back in time he sees.

How can we know the age of the universe? Perhaps the first clue is the age of the oldest rocks on the Earth. Using the technique of radioactive dating,[14] and employing the very long-lived isotope of uranium, the oldest-known rocks are found to be about four billion years old. The Earth must be older, because these solid rocks were once molten, as was the whole Earth.

Yet, the universe must be much older than the Earth for stars to have formed that create within their long lifetimes the elements from which rocks are made and the planets formed. How much older? To this question many avenues of inquiry lead to similar answers. Applying the radioactive dating

---

[14]In the molten fluid of the Earth, all sorts of elements were mixed together. When the Earth cooled and rocks solidified, they sealed for ages within them the radioactive elements and their decay products. It is just a matter of knowing their half-lives and some algebra to find how their proportions reveal the elapsed time between then and now.

technique to uranium itself is not as direct as the dating of rocks. But, by employing what has been learned about how stars forge the elements, an estimate can be made — somewhere between 10 and 20 *billion* years ago.

Other discoveries help pin down the age even more accurately. It is common knowledge nowadays that the universe is presently expanding, though this fact was not known at the time of Einstein's discovery and application of general relativity to the universe. The universe was thought to be static at the time (1919), just as Newton had believed.

The Nobel Prize was never awarded for this grandest of theories — general relativity. The term is familiar to everyone; it was not chosen by Einstein, and it does not provide any hint as to what the theory is about. General relativity is *the* theory of universal gravity. It is also more. Before Einstein, space and time were regarded as the stage, so to speak, on which things happened. Einstein made spacetime a part of physics in the deep sense that spacetime affects *and* is affected by what happens in the material world. This was a radically different concept from any theory that preceded Einstein's. It is remarkable that the theory was a purely intellectual achievement; it had no observational antecedent to suggest its need. The theory is regarded as the greatest accomplishment in theoretical physics, and it is due to one man — Einstein. He regarded the theory as being so beautiful and complete that he paid little attention when its first prediction was confirmed — the bending of light from distant stars by the Sun.

Einstein, himself, calculated the deflection, which is twice the Newtonian value. It was measured by Sir Arthur Eddington to be so. Eddington was among the very few who immediately understood Einstein's theory and grasped its scope. He organized and led an expedition to the island of Principe, off the western coast of Africa, at a longitude where an eclipse could best be observed and measurements made on the bending of light. He sent another expedition of British scientists to a location of similar longitude in Brazil. The results were reported to an excited gathering of the Royal Society in London shortly later. The news aroused great popular interest, being reported on the front pages of the major newspapers in Europe and America.

The universal expansion, which Einstein could have predicted but failed to, was actually discovered by Edwin Hubble at the Hale observatory in southern California. He was unaware of the theoretical work of the Belgian priest and cosmologist Georges Lemaître several years earlier in 1929. Lemaître envisioned and calculated models of an expanding universe, even an accelerated expansion, such as has now been observed.

Using the 100-inch telescope on Mount Wilson in southern California, Edwin Hubble (1889–1953) measured what is called the redshift, of light coming from distant galaxies. The redshift, also called the Doppler shift, is well known to anyone who has heard the high pitch of the whistle of an approaching train sink to a lower pitch as it passes. All wavelike motion will experience this sort of shift if it originates from a moving source. The pitch of a whistle, whether high or low, is determined by a wave property, namely the wavelength — the distance between adjacent crests of the wave. The wavelength of sound from the approaching train is shortened — its frequency increased — so that we hear a high pitch on approach.

Likewise, light has wavelike properties. Light from a star that is receding will appear redder: it is said to be redshifted. This is the Doppler shift of light.[15] The cosmological redshift, denoted by z, is the relative change in wavelength of light caused by the recession of the source, usually a galaxy. For small z, the redshift is related to the recession velocity by $v = cz$, where c is the speed of light (see Box 2).

The white light coming from a star is actually composed of all the colors of the spectrum, as Newton discovered when he passed light through a glass prism. The prism spreads the colors apart because each wavelength is refracted by a different amount, as in Figure 1.14. By measuring the shift of known spectral lines from a star with the same lines from a source in the laboratory, the astronomer can tell whether the star is moving away from or toward the Earth and at what speed. If the lines are shifted toward the red, he knows that the star is moving away. He can also identify the elements that are present in the distant source.

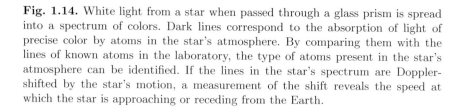

Fig. 1.14. White light from a star when passed through a glass prism is spread into a spectrum of colors. Dark lines correspond to the absorption of light of precise color by atoms in the star's atmosphere. By comparing them with the lines of known atoms in the laboratory, the type of atoms present in the star's atmosphere can be identified. If the lines in the star's spectrum are Doppler-shifted by the star's motion, a measurement of the shift reveals the speed at which the star is approaching or receding from the Earth.

---

[15]Sound and light each have their propagation speed. This does not change. But an approaching train compresses the distance between successive peaks of the sound wave, thereby increasing the frequency at which the waves arrive. Wavelength and frequency are seen to be inverse to each other.

Light seems to us in our everyday experience to be a continuous beam. And particles seem to be particles — quite discrete and distinct objects. However, light has some particlelike properties and particles have some wavelike properties. One way in which the particulate nature of light shows up is by the fact that its energy is carried in small bundles, the so-called *quantum* of energy. The energy carried by the quanta of light of one particular color is different from that carried by the quanta of another color. There is a continuous range of energy that light quanta can carry, but each corresponds to a different hue. That is why white light is spread into a *continuous* spectrum of hues when it is passed through a prism.

Similarly, the energy of motion of an electron in an atom is also quantized. Such an electron can absorb a light quantum, but only of just the precise energy that is needed to lift it from one quantized energy state of the atom to a higher one. Each type of atom — say, hydrogen, carbon, or oxygen — has a distinct pattern of energy states; the pattern is shown in Figure 1.15 for the simplest atom — hydrogen — which has one proton as its nucleus and one electron in orbit. Hence the energies of the light quanta that each type of atom can absorb are unique to that type of atom. According to the atoms that are present in the atmosphere of a distant star, the spectrum of light that arrives on the Earth has narrow bands of color missing, as in Figure 1.14. Such spectra are called *absorption* line spectra. By comparing the absorption lines in the spectrum coming from the distant star with lines produced by *known* atoms in the laboratory, the elements in the star's atmosphere can be identified. More than this, if the star is moving away from or toward us, each line in the pattern of absorption lines will be

Fig. 1.15. Quantum states (levels) of the *single-electron* hydrogen atom are represented by horizontal lines. Distances between levels represent the energy differences. Energies, as measured in a unit called the electron volt, are listed at the side.

shifted by a precise amount according to the speed of the star relative to the Earth. The Doppler shift is a tool of great value in astronomy.

1.4   Expanding Universe

> We find them [the galaxies] smaller and fainter, in constantly increasing numbers, and we know that we are reaching into space, farther and farther, until, with the faintest nebulae that can be detected with the greatest telescopes, we arrive at the frontier of the known universe.
>
> — Edwin Hubble

Edwin Hubble discovered in 1929 that other galaxies are moving away from us, as if the universe were expanding. He found that the light from distant galaxies was redshifted. This discovery profoundly changed our view of the universe. It was a great surprise to Einstein and most everyone, who thought that the universe was static. Moreover, Hubble discovered that galaxies at great distance are moving away more rapidly than nearby ones; quantitatively he discovered that the speed v of recession is proportional to the distance d of the galaxy, $v = H_0 d$, where H_0 is called Hubble's constant. This relationship is known as Hubble's law.

Hubble made these discoveries in the course of his attempt to determine the distances of celestial objects. First he measured the redshift of galaxies whose distances he already knew by other means.[16] They were found to obey the above relationship. This being so, the law could be applied to discover the distance to even more distant galaxies. Thus, by measuring the speed at which a galaxy is receding from us by means of the Doppler shift of light, the distance of the galaxy is given by Hubble's law. Many crosschecks verify these findings.

The value of Hubble's parameter that is measured at this particular epoch — that is to say, in our time — may be different than its value at other epochs because the pull of gravity of the universe may be causing the cosmic expansion to decelerate, or despite the pull of gravity, the expansion may be accelerating, pushed by a form of dark energy about which we will say much more later. The present value of the constant is known by the symbol H_0. The value of H at other times in the history of the universe is governed by Einstein's theory of gravity, as we will shortly see.

---

[16]Such as parallax, which we ourselves use in judging distance based on the two slightly different views that our two eyes provide.

Fig. 1.16. Edwin Hubble discovered that the universe is expanding. The law that is named after him states that distant galaxies recede from us at a speed that is in direct proportion to their distance. This and Hubble's other discovery — that the universe is uniform and similar in every direction we look — are key observations that underpin Big Bang cosmology. *Credit: Observatories of the Carnegie Institution of Washington.*

By measuring the Doppler shift, astronomers have learned that the most distant *visible* galaxies are moving away from us at speeds exceeding 1/2 the speed of light, and that their distances are about 13–14 billion light years.[17] Light that arrives on the Earth *now*, left the most distant stars that many billions of years *ago*. We see *now* how the distant stars appeared *then*. Meanwhile, by the time we have seen them, they have been continuing their journey. As to the age of the universe, it must be older than the time that light took to arrive at us from those early stars because the universe must have been older still for those stars to have formed out of tenuous gas clouds.

The Hubble constant, not the original value derived by Hubble, but the consensus value obtained by many types of observation in the intervening

---

[17]The most distant galaxies yet seen are at 13.4 billion light years. (See caption to Figure 6.4.)

years, provides the approximate age of the universe. The length of time that any pair of distant galaxies has been moving apart is given approximately by their separation divided by the speed at which they are moving apart. By definition this is the inverse of Hubble's constant $(t = d/v = 1/H_0)$.[18] The age determined in this way is about 15 billion years, and as a round number this is what we quote as the age of the universe (see Box 4).[19] That is the time in the past when the material that constitutes the stars today began its outward journey into the future.

As to the size of the universe that is *visible* to us, it is at least 13.4 billion *light years*, or, equivalently, 9.5 thousand billion kilometers.[20] It is not impossible to see to greater distance. Rather, up to that time there were few or no galaxies to shine. The universe at that time was filled only with enormous clouds of hydrogen gas and 25% helium. These gas clouds, a pale uniform light, and ghostly neutrinos that would travel almost unfelt and unfeeling in the cosmos for eternity, were all that existed then. However, their light was too dim to see.

But, is the distance that astronomers measure from their location here on the Earth really measuring the size of the universe? Might not some astronomer on the periphery of the universe as we see it, also see beyond us to such a distance? And so on. What then *is* the size of the universe? Is it even bounded?

1.5   Cosmological Principle

Over the ages, philosopher–scientists have been led inexorably to the realization that the Earth is not the center of the universe, or the Sun, or our own galaxy. The accumulating evidence has forced us to look ever outward. Peering deep into space, the view is similar no matter which direction we look. Galaxies, which are collections of billions of stars, are moving away from us; the greater their distance, the greater their speed. Presumably, our position in the universe is not special; an observer on the planet of a distant star will see a similar universe to the one we see. This assumption is known as the *cosmological principle* — we do not occupy a special place in the

---

[18]Here t is the recession time, d the separation, v the speed of recession, and H_0 the present value of Hubble's parameter. It has units of inverse time.

[19]The most recent data, as of this writing, suggest that the universe is 13.7 billion years old.

[20]A light year is the distance light travels in a year, about 10^{13} kilometers.

universe, but rather a typical one. At any given time, the universe would appear the *same* to any observer located *anywhere* in the universe. This is referred to as the homogeneity and isotropy of the universe. So not only are all distant galaxies rushing away from us, but they are all rushing away from *each other*. Where then is the center of the universe? *The universe has no center.*

1.6 Inflationary Epoch

The cosmological principle is logically compelling. We do not occupy a special place in the universe, but rather a typical one. Observers located in separate galaxies each see distant galaxies rushing away from their own location. Projecting backward in time, light from one observer will not have had enough time to reach the other. What made the universe so alike at all places even though no mechanism could have acted to make them alike at such an early time? An answer to this was proposed by Alan Guth. He supposed that early in its evolution (less than 10^{-33} s) the universe went through a period of extraordinarily rapid expansion. Prior to this, all parts of the visible universe could influence the other parts and they were homogeneous. There have been refinements to the original hypothesis. However, in this book we deal with the history of the universe beginning at a somewhat later and cooler time when the laws of particle physics would have gelled. The time at which the universe passed through that temperature was many orders of magnitude later than the inflationary epoch.

1.7 Beyond the Visible Universe

> Man is equally incapable of seeing the nothingness from which he emerges and the infinity in which he is engulfed.
>
> — Blaise Pascal, *Pensées*

Imagine two planets orbiting their suns, and these suns themselves lying in opposite directions across the wide expanse of the universe that is visible to us. As we found above, each is 13.4 billion light years away from us, and therefore 26.8 billion light years from each other. But the universe is only approximately 15 billion years old. Therefore light has not had sufficient time to cross the intervening space. Observers on one planet would have no possibility of seeing the other planet or star or the galaxy that they are in.

There hasn't been enough time since the beginning for light to travel that far. More than 11 billion years in the future the first light from one planet will be arriving at the other.[21,22]

The same for us. There is a limit to the distance that we have any possibility of seeing at this point in time and that distance is the age of the universe times the velocity of light, namely 15 billion light years. In the future we will be able to see further because the universe will have become older, and light will have had that much more time to travel to us from more distant reaches. The distance limit at any time is called the *cosmic horizon*. Beyond that distant horizon there are myriad other galaxies, stars and, no doubt, planets. And any observer who may be alive on one of those planets sees a universe very much like the one we see.

Is the entire universe finite or infinite; does it have a boundary or is it unbounded; will it last for a finite time or will it live forever? These are as-yet-unsettled issues, but astronomers are poised to unravel these wondrous mysteries.

1.8   Boxes 1–3

---

1   The Moon Falls Around

Make a sketch showing the Earth and the Moon's orbit; place a straight arrow originating at the Moon to indicate the *momentary* direction of travel, which is perpendicular to the line joining the Earth and the Moon. (For simplicity, use a circular orbit.) If the Moon is not to fly off into space in the direction of the arrow, note that it must be deflected from the direction indicated by the arrow's point by falling toward the Earth. It is forever falling toward the Earth but not reaching it because of its tangential motion. The tangential motion must have originated in the distant past during the formation of the solar system.

---

[21] Because of the universal expansion the two galaxies are moving away from each other even as light from one to the other is in transit.

[22] This paragraph points out a paradox having to do with the fact that galaxies, though too distant to have ever been in the same causally connected environment in the past, are nevertheless similar. Why and how could conditions have been similar instead of different? The paradox is among the motivations behind the theory of cosmic inflation.

2 Redshift

The redshift (also called the Doppler shift) of light emitted by a receding source — say, a galaxy — is the fractional change in the wavelength of light between that received by an observer λ_o and that emitted by the source λ:

$$z = (\lambda_o - \lambda)/\lambda.$$

Because of the scaling of wavelength with expansion ($\lambda \sim R$), it follows that

$$z = (R_o - R)/R \equiv \Delta R/R = (\Delta R/\Delta T R)\Delta T$$
$$= H\Delta T.$$

Distance to source is $R = c\Delta T$, while the Hubble law gives $v = HR$. So $z = (v/R)(R/c)$ or $z = v/c$.

3 Units and Data

$c \approx 3 \times 10^5$ km/s

$G \approx 6.7 \times 10^{-8}$ cm^3/gs^2

Year $\approx 3.2 \times 10^7$ s

Distance: ly (light year) $= 9.5 \times 10^{12}$ km

Distance: pc (parsec) $= 3.3$ ly $= 3.1 \times 10^{13}$ km

Hubble constant: $H_0 \approx 200$ km/s $\cdot 1/(10^7$ ly$)$

$1/H_0 \approx 15 \times 10^9$ y $\approx 4.6 \times 10^{17}$ s

Parsec ≈ 3.3 ly $\approx 3.1 \times 10^{13}$ km

Universe's age $\approx 1/H_0$

Earth's age $\approx 4.5 \times 10^9$ y

Earth's mass $\approx 6 \times 10^{27}$ g

Sun's age $\approx 4.5 \times 10^9$ y

Sun's mass $M_\odot \approx 2.0 \times 10^{33}$ g

Milky Way's age $\approx 10^{10}$ y

Milky Way's mass $\leq 10^{12} M_\odot$

Life of a $10 M_\odot$ star $\approx 10^7$ y

Time since dinosaurs $\approx 7 \times 10^7$ y

"$\approx$" means accurate to two figures

2 The Very Large and the Very Small

What is it that breathes fire into the equations and makes a universe for them to describe?

— Stephen Hawking, *A Brief History of Time*

2.1 Face to Face

The very word "cosmos" inclines our minds and our imaginations toward the infinite. The cosmos is indeed very large and is becoming ever larger. At the same time the behavior of its constituents on the scale of atoms and their nuclei can be understood only at the other extreme. Thus, relativity and quantum mechanics come face to face.

2.2 Space, Time, and Relativity

There is no absolute relation in space, and no absolute relation in time between two events, but there is an absolute relation in space and time.

— A. Einstein, *The Meaning of Relativity*

In our normal everyday experience, or even in the broader field of technology, like rockets and spacecraft, speeds are small compared to the maximum possible speed. And, according to our understanding of gravity, the gravitational forces exerted on objects on the Earth are small compared to those at the surfaces of neutron stars or at the horizons of black holes. Therefore, it should come as no great surprise that beyond the realm of our own experience there are processes both here on the Earth and in the universe that defy our intuition.

Time seems to flow like a river at a constant rate past us. However much we may wish to speed it up, or slow it down, it seems quite beyond all power to do so. Space too, like time, seems absolute. Surely, we think, two identical rulers that exhibit the same length when set side by side will always appear to be identical. These were Newton's commonsense views of time and space, and they seem to be in accord with our everyday experience. *Things* happen *within* absolute space and absolute time but *they* do not affect space and time. So it seems to us.

Einstein (1879–1955) had ideas about time and space that were very different. He viewed time and space as *part* of physics, and thus as being subject to physical laws. To compare times on clocks in two moving systems he needed a messenger and he used light, which he postulated is constant as viewed by all observers, even though they themselves may be moving with respect to each other. Strange as this may seem, it was borne out by experiment (Michelson-Morley).[1] Second, he postulated the principle of relativity—that the laws of physics will be measured to be the same in laboratories that are moving uniformly with respect to each other.

It is remarkable that such a beautiful theory as special relativity (1905), with such far-reaching consequences, is founded on such seemingly *unremarkable* postulates. However, for Albert Einstein they were pregnant with a deep meaning — space and time do not have a separate and absolute meaning, but *spacetime* does. Two consequences of his postulates that are quite startling but well known and have been proven beyond doubt are the contraction of space and the dilation of time as measured by observers in uniform motion with respect to each other.

Let us elaborate two special effects of relativity spacetime. Imagine two people carrying identical clocks which, standing side by side, they adjust to have precisely the same time. And they also have sticks, which they carve to have identical lengths. Each person takes off in a separate rocket ship. After coming to speed, they turn about and zoom past each other while holding their sticks in line with the direction of flight and the clocks in hand so each can see the other's. Both experimenters carefully measure the length of the other's stick, and note the time on the other's clock. Each one finds that his companion's stick measures shorter than his own, and the time between ticks of his companion's clock is longer than that between ticks of his own clock. They have discovered the dilation of time and contraction of distance for observers in uniform motion at constant speed with respect to each other.

Of course, when speeds are small compared to that of light, Einstein's special theory reduces to Newton's in the limit. This is as it must be because Newton's theory describes mechanics very well under the usual

---

[1] Einstein appears not to have been moved by the experiments, if he knew about them at the time. A. *Pais*, Subtle Is the Lord, *Oxford University Press*.

circumstances of everyday life. Neither theory works for strong gravitational fields, as Einstein understood, practically from the outset. This realization drove him to years of hard work in searching for a theory of gravity, often called *general relativity*.

The predictions of Einstein's special theory of relativity are quantitatively employed as a matter of routine in every high energy particle laboratory around the world. For example, the dilation (or elongation) of time has been used to discover very unusual particles whose lifetimes are so short that they do not exist naturally in today's universe. By supplying the energy equivalent of their mass and more in particle accelerators they can be produced with velocities nearly equal to the velocity of light. Then they live long enough — as observed from the laboratory — to be detected and for their dilated lifetime to be measured. The particle's own lifetime, t, is related to the dilated time, T, observed in the laboratory relative to which the particle is moving, by the relation

$$T = \frac{t}{\sqrt{1 - (v/c)^2}} \, .$$

Why is it important to learn about unusual particles that no longer exist naturally? Because the early universe whose high temperature afforded the energy needed to create them must have been pervaded by a great multitude of various particles and their antiparticles. The cosmologist needs this kind of information — particle types, their lifetimes, their interrelations— to trace how the universe evolved from near its beginning to the present universe of galaxies, stars, planets, and atoms of many kinds.

2.3   Physics of the Very Small: Quantum Mechanics

If astrophysics does not already draw on all branches of physics, it surely will. From the late 19th century to the 20th the foundations were laid for understanding the two extremes at which classical Newtonian physics failed—it failed for the the very fast and the very massive (Einstein's special and general relativity), and it failed for the very small (quantum mechanics).

Though we think of astronomy as dealing with very large objects like stars and galaxies, to really understand these objects and how they came to be, we need first to understand atomic, nuclear, and elementary particle physics. The laws of quantum mechanics govern such small objects, not the laws of ordinary mechanics which work so well in the world on the scale that we, as humans, experience it.

That something was seriously amiss with classical ideas was becoming ever more evident to a number of theoretical physicists in Europe around the beginning of the 20th century — Lorentz in Holland, Planck and Hilbert in Germany, Poincare in France, Fitzgerald in Ireland, and, of course, Einstein in Switzerland and Germany. The invention of quantum theory to understand microphysics was the work of many giants such as Planck, Einstein, Bohr, Pauli, de Broglie, Schrödinger, Heisenberg, and Dirac. Relativistic kinematics (special relativity, 1905) and the modern theory of gravity (general relativity, 1915) was developed by one man—Einstein.

Knowing Einstein's doubts about quantum theory, one may wonder that he was one of its founders. Indeed he was. His insight into the photon nature of light profoundly influenced the melding of particle–wave duality by de Broglie and the wave theory of Schrödinger. Indeed, Einstein won his Nobel Prize (1921) for this work. And his last judgment of quantum theory was that it is a logically consistent theory of great value; only that it is not a complete theory. Most theoretical physicists agree.

The laws of special relativity and quantum mechanics govern the very fast and the very small, respectively. These theories do not lend themselves to an easy understanding. Our intuition of nature is based upon our own experience of the macroworld, not the microworld. And the part of nature that we can experience with our senses and on which our intuition is based is a very limited part of the whole. These theories — relativity and quantum mechanics — do not disagree with the laws of physics that we experience in our everyday life. Rather, they blend smoothly into them.

An important aspect of the small-scale physical world is expressed by quantum mechanics in what is called *particle–wave duality*. Separately, particles and waves fall well within our everyday experience. At our experiential level, particles, like billiard balls, are well-localized objects that carry mass, energy, and momentum when they move. In contrast, waves, like ripples on a lake, are delocalized — they are spread out in space and carry energy *but not mass*. However, these distinctions are blurred at the small scale — but not completely. Particles have some wavelike properties; and waves, like light, have some particlelike properties; but neither has all the properties of waves *and* particles. For example, particles have mass and can move at any speed up to but *not* including that of light. Light has no mass and only one speed — yet, like a particle, it has momentum.

Certainly, light has a wavelike character. We know this from the double-split diffraction experiment (Figure 2.1). We also see wave interference on the surface of a quiet pond when the outward-moving concentric rings of waves from two different disturbances meet. The interference of wave with

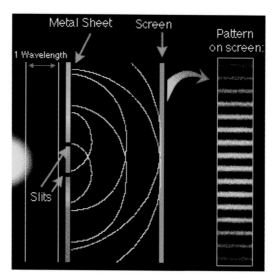

Fig. 2.1. Light from the bulb at the left reaches the parallel slits in the metal sheet. On the other side it is as if we were looking down into a pool of water when a circular wave emanating from the bulb at the left sets up two waves on the other side of the sheet. These waves interfere with each other, reinforcing the pattern at intersections of the tops of the waves and making deeper troughs at the intersections of the troughs. When the light falls on a screen as at the right, the pattern of reinforcement and cancellation shows up in what is called a diffraction pattern. *Credit: F.R. Spedalieri, NightLase Technologies.*

wave is quite different from the interference of billiard ball with billiard ball, as we witness on our scale of things. Nevertheless, very small particles like electrons behave similarly to light in particular circumstances. This has to do with the quantum nature of the microworld.

Einstein won his Nobel Prize in 1921 after having been nominated in 10 of the previous 12 years by the most eminent scientists of the day.[2] In the year he was finally awarded the prize, one of those who had made a nomination asked the Nobel committee to look 50 years into the future and imagine what would be thought then if Einstein had not been awarded the highest prize in intellectual achievement. He won it not for relativity — which in the opinion of the Nobel committee was not firmly enough established at that time — but "for his services to Theoretical Physics, and especially for his discovery of the law of the photoelectric effect". In fact, the Nobel Prize was never awarded for the most singular of all intellectual

---

[2]A. Pais, *Subtle Is the Lord* (a biography of Albert Einstein' scientific career).

achievements — general relativity, the very theory which enables us to formulate the theory of cosmic history.

In one of Einstein's three famous papers of 1905 (the other two were on the special theory of relativity and Brownian motion[3]), he made the audacious assertion that light consists of quantized bundles which he called *photons*.[4] With this theory, he was able to explain very puzzling experimental observations. It had been observed that when a ray of light of a fixed color fell on a smooth metal surface, electrons were emitted, and that all of them had the same energy. More puzzling, their energy did not increase when the intensity of the beam was increased. Instead, a greater number of electrons were dislodged. Only when the color was changed to one of a higher frequency (greater energy) did the electron energy increase. Einstein realized that all of this showed that the beam of light was behaving like a beam of particles by knocking electrons from the metal surface and, moreover, that a certain minimum energy was needed because each electron was held to the metal by the attractive electric force exerted by the protons in atoms of the metal. This phenomenon is called the photoelectric effect.

Prince Louis de Broglie took the next step in melding together the wave- and particlelike behavior of small-scale phenomena. As a youth, he entered the Sorbonne as a history student and graduated with an arts degree at the age of 18. By that time he was already becoming interested in mathematics and physics and in 1913 de Broglie was awarded a *Licence des Sciences*. However, before he made further progress, the First World War engulfed Europe, and the young de Broglie was attached to the wireless telegraphy section in the French Army. "When in 1920 I resumed my studies...what attracted me...to theoretical physics was...the mystery in which the structure of matter and of radiation was becoming more and more enveloped as the strange concept of the quantum, introduced by Planck in 1900...penetrated further into the whole of physics."

---

[3]Brownian motion concerns the observation that in an otherwise clear liquid, if a suspension of small particles is introduced, the particles appear to be knocked about as if being struck by other invisible particles; indeed they are. The other particles are the molecules of the liquid, illustrating that a liquid is not a smooth uniform medium but is composed of closely spaced molecules. Their motion becomes more agitated as the liquid is heated because they jostle the particles of the suspension more vigorously. All of this is obvious to us today, but was a step along the way to understanding the atomicity of matter.

[4]Max Planck had earlier introduced the constant called after him, h, as a necessary constant in his mathematical formula for the spectrum of radiation from a perfect emitter, and absorber, called in physics a *black body*.

Fig. 2.2. Louis de Broglie in Paris about the time of his discovery of the particle–wave duality for *particles*. *Credit: Fondation Louis de Broglie, 23 rue Marsolan, Paris.*

Einstein's interpretation of the photoelectric effect together with the emerging quantum theory convinced Louis de Broglie that both light waves *and* particles possessed a *wave–particle duality*. He developed the theory of *electrons as waves*, whose wavelength λ he related through Planck's constant h, the electron mass m, and its velocity v. He asserted that under some circumstances the electron would behave as if it were a wave with a wavelength $\lambda = h/mv$.[5] Legend has it that de Broglie's professors at the Sorbonne consulted Einstein about whether he should receive a degree for his thesis. Einstein replied that he had better be given a Nobel Prize. The theory was proven correct in experiments performed in 1927.[6] He was awarded the prize two years later.

One of the ways in which electrons behave as waves is the precise analogue of the double-split diffraction pattern of light (Figure 2.1). If a beam of

---

[5] *Comptes rendus de l'Acadmie des Sciences*, Vol. 177 (1923), pp. 507–10.

[6] C.J. Davisson, C.H. Kunsman, and L.H. Germer in the United States and G.P. Thomson in Scotland.

electrons is shone onto a thin plate with two very narrow and closely spaced parallel slits, a diffraction pattern is created on a photographic emulsion plate on the other side, just as for light. How can this be? Doesn't the electron have to go through one slit or the other, but not both? If one slit is covered, indeed, the emulsion plate will register all the hits in line with the slit and source of electrons. But the mere presence of the other opening altogether alters the outcome; an outcome that is the same as if a light source had been used instead of electrons. Think of that! The mere *possibility* that each electron could have gone through either slit affects what the electron actually does; and the pattern it makes on the photographic plate does not resemble at all the pattern that would have been made if first one slit were closed, and some time later the other.

Schrödinger was immediately fascinated by de Broglie's thesis. Shortly after reading it, he published his now famous paper in which he formulated the wave theory of quantum mechanics; the Schrödinger wave equation appears there for the first time. He was awarded the Nobel Prize in 1933 together with Dirac for this work.

2.4 Heisenberg's Uncertainty Principle

Werner Heisenberg (1901–1976) is best known to the public for a discovery that he made at the age of 26. That discovery concerned the behavior of particles at the atomic and subatomic level. Purely by theoretical reasoning he discovered the *uncertainty principle*, often referred to with his name preceding it. His statement was succinct: "The more precisely the position [of an atomic particle] is determined, the less precisely the momentum is known in this instant, and vice versa." This discovery lies at the very roots of quantum mechanics, and was the inspiration for Dirac's great leap forward in that field.

2.5 Radiation in the Early Universe

The early universe was filled with light, and light played a very important part in the way the universe evolved from its intensely hot beginning.[7] As we will come to see, light links the present to the past in several vital ways that

---

[7]In the biblical account of creation, the earth was formless and empty. Darkness was on the surface of the deep. God said, "Let there be light," and there was light. Cosmology deals not with the darkness before the beginning, but with the light and everything else after the moment of creation.

Fig. 2.3. Werner Heisenberg as a young man in 1927, the year when he published his famous paper on the *uncertainty principle* in quantum mechanics. This was among the half dozen major foundation works of that important field, the mechanics that govern the world at the atomic and subatomic level. © *The American Institute of Physics.*

have permitted the cosmologist to form a quite detailed description of how all that we see, came to be. Because light plays such a role, and in any case is interesting in many other ways, we pause here to describe its nature. Neither will it lead us too far astray to comment on the special relationships between the peak wavelengths of the Sun's radiation, the particular absorption and transmission properties of the Earth's atmosphere for radiation, and the size and spacing of light-sensitive cells in the retina of the eye. No doubt these relationships enabled and caused animal life *on this planet* to evolve in such a way as to use light — which is one small portion of a very broad spectrum of radiation — as its most important sensory link with the world.

At the very high temperatures of the early universe, *light* means much more than that small part of the spectrum of radiation — the optical spectrum — that we see with our eyes. Other parts of the spectrum with which we are familiar in everyday life range all the way from radio waves at one end to ultraviolet and X-rays at the other. But beyond these, the early cosmos was filled with radiation of which we have little conception. The cosmos was filled with *particle–antiparticle* pairs of many types, not only of electrons, neutrons, and protons from which atoms are made, but also many types of particles that do not normally occupy the world today. We know that these and their antiparticles can exist because experimenters, using high energy particle accelerators, have discovered that they can be created for a *short* time. And there may be still other unidentified particles that have not yet been discovered, but whatever they are, we know that they must have existed in the intense fire of the early universe because then the concentration of energy was even much higher than can be produced in laboratory machines.

How can we know that all such particles populated the early universe? As we will see in the next chapter, there is abundant evidence that the universe was intensely hot at the beginning. And heat is a form of energy. A part of that initial high concentration of energy must have existed as particles because energy can transform itself or be transformed into many forms, including the mass of particles ($E = mc^2$). Indeed, according to an old and well-established field of physics known as statistical physics, the energy must have been shared between the heat energy, which corresponds to the agitated motion of particles, and newly created particles and their antiparticles.

The word "antiparticle" may be widely known but not its exact meaning. We will encounter antiparticles again (Section 4.2, page 97) but, for now, note simply that they have properties that are just the opposite of particles. If one has a negative charge, the other has a positive charge. In this way their properties cancel each other. Just like a photon, the particle–antiparticle *pair* carries no charge; but like a photon the *pair* carries energy and momentum. So the pair is interchangeable with a photon. Photons of radiation can disappear and be replaced by particle–antiparticle pairs.

Conversely, when a particle and an antiparticle meet, and they did so frequently in the fiery cauldron that was the early universe, they annihilate each other but their energy and momentum can never be lost. Generally, a photon, the particulate form of light, emerges carrying off the conserved quantities. Particles, antiparticles, and photons collided frequently in the dense early universe on a timescale much shorter than that of the expansion of the early universe. All things were in a state of flux, but on the average maintained a certain order called thermodynamic equilibrium.

Our eyes are sensitive to a small range of electromagnetic radiation that we call light. We experience light in all the colors of the rainbow. This is one of the wonders performed by our nervous system on another wonder of the natural world — the spectrum of electromagnetic radiation. I say it is a wonder because the objective reality described above is so different from the subjective one. Light is only a small part of the spectrum of radiation. It is the *only* part that we "see" with our eyes. Through an amazingly complex chemistry involving solutions containing sodium and potassium ions in receptors in the retina of the eye, the energy of photons from the outside world is converted into electrical signals that the brain interprets in terms of form, movement, and color.[8] *Rods* in the retina are especially sensitive to dim light and transmit information only on the intensity of light, providing night vision in shades of gray. *Cones* in the retina are of three types that absorb strongly at three different wavelength bands. The differences in absorption between these bands provide the sensation of color. The external reality is merely a matter of wavelength of the electromagnetic radiation emitted or reflected by an object in the external world. The sensation of color is purely a creation of the nervous system. What a marvel the world is in every aspect!

Another part of the radiation spectrum which we sense, not through our eyes but through our skin, is the radiant heat of the Sun. A third is the ultraviolet spectrum that we neither see as light nor feel as heat, but which is powerful enough to destroy the skin if exposed too long. Others that we sense indirectly are radio waves that a receiver translates for us into vibrations in the air, which we hear as sound.

All of these radiations from radio waves to ultraviolet and beyond have the same nature: they are produced by the oscillating motion of charged particles such as electrons.[9] If the *frequency* of the oscillation is slow (a million times a second), radio waves are produced; if moderate (trillions per second), infrared and optical (light) waves; and if very rapid (millions of billions per second), ultraviolet, X-rays, or gamma rays are produced. From Newton's laws of mechanics, force is required to accelerate a particle

---

[8] J.L. Schnapf and D.A. Baylor, "How photoreceptor cells respond to light", *Scientific American*, Vol. 256 (1987), pp. 40–7.

[9] Oscillation of an electron involves a rhythmic pattern of acceleration and deceleration and therefore produces a flow of photons of the same frequency or energy. But any acceleration (deceleration) of a charged particle produces photons of radiation.

$(F = ma)$ and force applied to the particle over distance requires energy $(E = Fd)$. Energy is conserved. The energy that is expended by whatever natural or man-made machine that caused the acceleration of the charged particle has been transformed into energy that is carried by the radiation. And that energy is carried outward from the source — the oscillating charge — at the speed of light, though it could as well be called the speed of X-rays, or any massless particle. All of these radiations travel at the speed c.

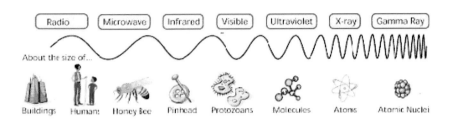

Fig. 2.4. The wavelengths of electromagnetic radiation span lengths of the size of city blocks (radio waves) to the size of protons (gamma rays). Visible light occupies one small part, or band, of the spectrum, as shown. *Credit: National Aeronautic and Space Administration.*

The wavelengths of the radiation spectrum span a vast range of dimensions, from the length of city blocks at the one extreme to molecular, atomic, and nuclear dimensions at the other (Figure 2.4). The longest wavelengths are radio, followed by infrared that we experience as warmth from the Sun or a stove. The visible or optical wavelengths are around $1/20\,000$ centimeters and constitute only a minuscule part of the band of radiation. Still shorter wavelength radiation is known as ultraviolet, X-rays, and gamma rays and their wavelengths range from $1/10\,000\,000$ centimeters down to $1/10\,000\,000\,000\,000$ centimeters. When the universe was very young and small, only very short wavelength radiation could fit into it; that is to say, high energy gamma rays.

In the hot plasma of the early universe, all particles were in rapid motion and frequently encountered one another. They exerted a force on each other through the interaction of their charges — they attracted or repelled one another, depending on whether the signs of their charges were alike or unlike. Force causes acceleration $(F = ma)$, and the acceleration of the electric field of the charged particles can create a quantum or photon of radiation. The converse can also occur; the energy and momentum of a photon can be partially or totally absorbed by a charged particle. By means of frequent interactions of these two types — *inelastic scattering*, as they are called — the photons and the particles share energy in particular ways that are governed by laws of statistical mechanics. Through these processes, a state of *thermal* equilibrium is reached. Although the individual particles and photons frequently interact and have their energy changed, the average way the energy is shared remains constant, according to the temperature and the Bose–Einstein or Fermi–Dirac laws.[10] Temperature measures the average energy of random motion of all the particles, antiparticles, and photons in thermal equilibrium.

There is another, very important way in which electromagnetic radiation can be created, modified, or destroyed. It is through its interaction with atoms, molecules, or nuclei. These entities contain charged particles so that they can interact with radiation. The motion and internal structure of atoms, molecules, and nuclei are also governed by the laws of quantum mechanics. As a consequence, the amount of internal energy contained by, say, an atom, cannot have any arbitrary value but only one of a number of discrete values. The normal state or lowest energy state of an atom is called its ground state. Other quantum states of the atom are called excited states, and they differ in energy by discrete amounts. This is illustrated in Figure 1.15 for the simplest (but very important) atom, hydrogen. Every type of atom has its own *particular* pattern of energy levels.

Hydrogen has one proton as its nucleus and is orbited by one electron. A photon can be absorbed by an atom only if it has just the right energy to raise the electron from its lowest energy or ground state to one of the higher states. It ceases to exist, and the electron in an excited state of the atom carries its energy. The mass of the atom is actually greater after absorption by an amount equal to the energy of the photon, in accord with Einstein's equivalence of mass and energy, $E = mc^2$. Some time later the excited electron can emit a photon of just the amount of energy by which it differs from a lower state. The two processes of absorption and emission were very important in the plasma that occupied the early universe. They are additional ways in which energy can be transmitted and shared between radiation and particles, the electrons, protons, nuclei, atoms, and molecules.

The Earth's atmosphere affects very strongly what radiation can reach its surface (Figure 2.5). The atmosphere contains a mixture of a number

---

[10]Particles at the atomic level carry *spin* which can have either an integer $(0, 1, \ldots)$ or a half-odd integer $(1/2, 3/2, \ldots)$ in units of Plancks constant, $h/2\pi$; such categories of particles are known as bosons and fermions respectively.

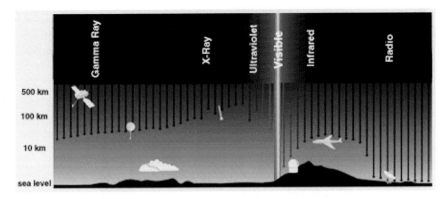

Fig. 2.5. The depth of penetration through the Earth's atmosphere of various wavelengths of radiation from the Sun is shown by vertical lines extending from the top of the figure. The atmosphere is transparent to radiation in the visible range of the spectrum and fairly transparent in the infrared. The Sun's luminosity is peaked in the visible band. For these two reasons, the Sun provides light and heat to life on the Earth. Fortunately, for life the atmosphere is opaque to X-rays, gamma rays, and ultraviolet. But, because of the opacity, astronomers who seek information about the universe by studying radiation at these wavelengths have to fly their detection instruments in balloons or satellites. On the other hand, radio astronomy can be conducted day and night with Earth-based telescopes.

of gases (atoms or molecules of various types); it is most dense at the surface, and becomes ever thinner with altitude, extending to a height of hundreds of kilometers. As discussed above, atoms, molecules, and nuclei can absorb photons. They also re-emit photons, but not necessarily of the same energy as that of the one absorbed. Rather, a whole sequence of lower energy photons may be emitted instead, each corresponding to the energy difference between excited states of the atom, molecule, or nucleus. Therefore, the particular atmosphere around the Earth interacts with the various wavelengths of radiation coming from the cosmos in different ways. It is transparent to light, to which our eyes are sensitive. This means that light in the optical range easily reaches the Earth's surface. This is fortunate for animals that use light to see, and for astronomers who use very large optical telescopes. Still, there is some absorption (and scattering), which is why telescopes are usually built on mountaintops, and, more recently, carried into orbit above the Earth's atmosphere. The longer wavelengths of infrared radiation also reach the Earth's surface and warm it. On the other hand, high energy radiation (very short wavelength) is strongly absorbed in the atmosphere (by atoms and nuclei). Therefore, X-ray and gamma

Fig. 2.6. An artist's impression of a distant star (yellow) orbited by a planet and its moon (foreground). ©*David A. Hardy/www.astroart*

ray astronomy is done almost exclusively with balloon- and satellite-based detectors. High energy radiation would be very damaging to life, so it is indeed fortunate that the Earth's atmosphere is opaque to it.

2.6   Other Planets Around Other Suns

> There are infinite worlds both like and unlike this world of ours. For the atoms being infinite in number, as was already proven, ... there nowhere exists an obstacle to the infinite number of worlds.
>
> — Epicurus (341–270 B.C.)

Since we first learned about the immensity of our galaxy, containing as it does 400 billion stars, and with light taking 90 000 years to cross it, we have wondered if there are planets around other stars, and whether there is life on any of them. The first planets outside our solar system were discovered in 1992 by means of the Doppler shift, the same way in which Hubble discovered that the universe is expanding.[11] But those three planets orbit a neutron star, which, lacking light and heat and bombarded by high energy radiation, would be very inhospitable to life.

Are there planets around other suns? Two groups — one in Switzerland led by M. Mayor and D. Naef, and one in the United States by G.W.

---

[11] A. Wolszczan and D.A. Frail, *Nature*, Vol. 355 (1992), p. 145; A. Wolszczan, *Science*, Vol. 264 (1994), p. 538; A. Wolszczan, *Science*, Vol. 264 (1994), p. 538.

Fig. 2.7. Geoff Marcy, codiscoverer of more than 100 planets in our Milky Way galaxy which lie far outside the solar system. *With permission of G.W. Marcy.*

Marcy and R.P. Butler — took up the challenge of such a search a few years ago. Marcy's group has found 110 of them (as of September 2003) lying *outside* our solar system and in the Milky Way (Figure 2.7). The length of their *years* ranges from 7 hours to 7 years. Their masses range widely in comparison to the planets in our own solar system—between 0.3 and 16 Jupiter masses.

Even with the most powerful telescopes, a planet orbiting a star that is outside our solar system cannot itself be seen. Rather, the gravitational pull on its sun can be detected as a wobble of that sun if the planet is sufficiently massive and that sun is not too distant.[12] When the planet lies between its sun and us, the planet pulls its sun toward us through their mutual gravitational attraction; when the planet lies on the other side, it pulls its sun away from us. The light of the sun is therefore alternately Doppler-shifted toward the blue end of the visible spectrum and then toward the red in a sinusoidal pattern. This is entirely analogous to the Doppler shift in the pitch of a train whistle, which rises or falls according to whether the train is approaching or receding. The earthly astronomer is able to detect the planet's presence by this slight shift in the color of that star.

If the sun that is under observation has more than one planet, the pattern of the changing color is more complicated, but by careful analysis,

---

[12]G.W. Marcy and R.P. Butler, *Astrophysical Journal Letters*, Vol. 464 (1996), L147.

called a Fourier analysis after the mathematician of that name, the separate effects of each planet can generally be untangled, especially for massive planets. But the Earth's mass is about 1/1000 that of Jupiter's and no planet resembling the Earth has yet been found. This does not mean that such planets do not exist, but only that their pull on their sun is too small to be seen by present instruments.

From *observation* it has been estimated that at least 10% of all stars in the Milky Way have large planets the size of Jupiter and Saturn. From the 400 billion stars in our galaxy, this implies a total of 20 billion large planets in the Milky Way. From the *theory* of planet formation and our existing detections of giant planets, it is estimated that actually 50% of all stars have terrestrial planets, implying a total of 100 billion in the Milky Way alone.[13]

2.7   Life on Other Planets

Are any of these planets habitable? Conditions for very primitive forms of life are not very demanding. Primitive life permeates the Earth from the highest coldest mountain to the deepest sea, and even deep within the Earth. We do not know how life emerged from the inorganic world. Nevertheless, given the vast number of primitive forms, it seems not improbable that primitive life exists on some of the other planets, possibly even in our solar system.

However, conditions for higher forms of animal life are very demanding. One condition, especially for most mobile animals, is sensitivity to light as a means of navigation (Figure 2.8). Sound would do, but evidently is not as versatile a means as light. As we have discussed, light constitutes an extremely small part of the electromagnetic spectrum, as Figure 2.4 so vividly displays. Could other parts of the spectrum be used instead? The answer is no. To be sensitive, to radiation in the sense of seeing, the cellular structure of the retina, the rods and cones, must be about as small and as closely spaced as the wavelength of the radiation to which an eye is sensitive, as depicted in Figure 2.9. Clearly the minimum size of biological cells of any sort is determined by their function and the size of complex organic molecules. Short wavelengths, such as ultraviolet or X-rays (high energy photons), would not be suitable because they destroy tissue. Longer ones would hardly provide enough energy per photon to activate electrical signals needed for transport of information to a brain; moreover, the information that longer wavelengths could convey, if any, would be very coarse. (The

---

[13]Estimates courtesy of G.W. Marcy, University of California, Berkeley.

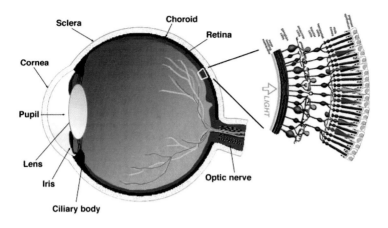

Fig. 2.8. Human eye showing enlargement of nerve structure in the retina. Our eyes evolved to be sensitive to the optical range of wavelengths because living cells can be organized into receptors having such dimensions as these wavelengths. H. Kolb (1991), in *Principles and Practices of Clinical Electrophysiology of Vision* (Eds. J.R. Heckenlively and G.B. Arden), Year Book Inc., St. Louis. *Credit: Helga Kolb, Eduardo Fernandez, and Ralph Nelson.* Many images can be viewed on http://webvision.med.utah.edu/index.html

human eye can detect single photons.[14]) Therefore it seems fairly certain that the small range of radiation at *optical* wavelengths is the only range that could be useful for vision.

How common would it be that a star radiates appreciable energy at optical wavelengths? Not very. Stars vary greatly in temperature, according to their mass and age. Figure 2.10 shows the spectrum of radiation of three stars of different temperature — our Sun and one hotter, one colder. The hotter one would provide little optical radiation and even less heat (infrared) in comparison with the extreme destruction that the high energy radiation would wreak if life were close enough to benefit from the useful radiation. The colder star would provide too much heat for life on a planet that was close enough for the lower intensity of the optical radiation to be useful for vision.

Finally, there are special conditions that the planet itself must provide if higher life forms — especially those that are familiar to us on this planet — are to inhabit it. First, the planet's sun must be stable in size and

---

[14]D.A. Baylor, T.D. Lamb, and K.W. Yau, "Response of retinal rods to single photons", *Journal of Physiology, London*, Vol. 288 (1979), p. 613.

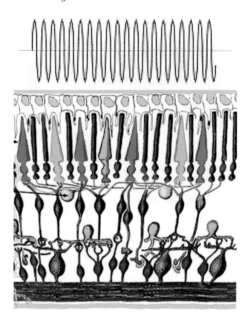

Fig. 2.9. I have adjusted the wave form (top panel) representing visible light to correspond to the dimensions of the narrow range to which the retinal sensing circuitry of the human eye (bottom panel) has been adapted. The correspondence in the spacing and size of rods and cones is clearly visible in the circuitry so as to provide the brain with information that can be interpreted as shape and color. *Credit: Helga Kolb, Eduardo Fernandez, and Ralph Nelson.*

energy output, and the planet's orbit must be close to circular so that the seasons are not too different. The planet's atmosphere must have certain properties. Aside from having oxygen, carbon, and water the atmosphere must be transparent to some radiation and opaque to others. Transmission of the Earth's atmosphere is illustrated in Figure 2.5. Light passes easily through the Earth's atmosphere, as does the near infrared (heat). At the same time the atmosphere shields life on the Earth's surface from harmful X-rays and gamma rays. The Earth's atmosphere is especially friendly to seeing animals.

Is there life on other planets, especially higher forms? It seems likely, given the estimate that 100 billion stars in the Milky Way have planets of their own. Still, we do not know what circumstances brought forth life from the inanimate world on our own planet, nor how likely those conditions have occurred on others. And certainly the distance — even of the nearest stars,

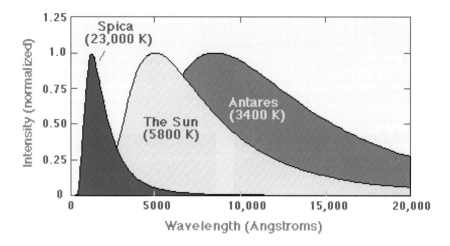

Fig. 2.10. The Sun's spectrum of radiations is peaked in the visible and strong in the near infrared, and because the Earth's atmosphere is transparent, or relatively so at these wavelengths (see Figure 2.5), the Sun warms the Earth and provides light that is absorbed by plants and converted to food by photosynthesis, furnishing all animal life with food. The other two stars, Spica and Antares, would be very inhospitable to life on any planet that might orbit them. The one would provide no light but very strong high energy radiation. The other would provide little light but much heat. Surface temperatures are shown in parentheses. *Used with permission,* Online Journey Through Astronomy, *M.W. Guidry, Brooks/Cole Publisher.*

Proxima Centauri and Alpha Centauri — is so great that it would take nine years to receive an answer to a message from the Earth.

2.8   Different Points of View

Our eyes are sensitive only to a very narrow band of the broad range of wavelengths of radiation that can be produced under various conditions in the universe. However, with other instruments and technologies, the astronomer can detect objects and processes in the universe that produce radiation across the entire band. Because of the particular transmission properties of the Earth's atmosphere, some of these instruments must be borne aloft in balloons or satellites.

Wavelengths of radiation that are not visible to us, such as infrared or gamma rays, can nonetheless be visualized in what are called *false color images.* By mapping invisible wavelengths onto arbitrarily chosen colors of the visible spectrum, an impression of the shape of the emitting object and the intensity of radiation from its various regions can be obtained (Figure 2.11). Still, this is not enough for the astronomer who will want to measure the actual wavelengths and their intensities, and possibly variations in intensities over time; the study of astronomy goes much deeper than pictures like Figure 2.11 alone can reveal.

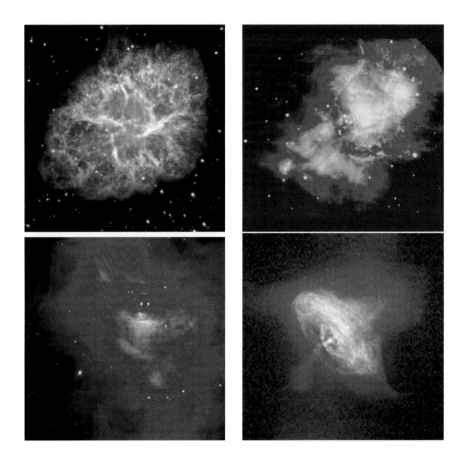

Fig. 2.11. The Crab Nebula, which is a supernova remnant, is now about 10 ly across and is expanding at 1800 km/s. It is seen here in four different views. The bottom pair are magnified views of the region around the Crab neutron star rotating 33 times per second. Wavelets can be seen emanating from the region of the neutron star from which back-to-back jets can be seen. *Credit: Infrared — W.M. Keck Observatory; optical — Mount Palomar; neutron star region (optical) — NASA; X-ray — NASA/CXC/SAO.*

The Crab Nebula is the shattered remains of a star whose demise was observed by the Chinese astronomer Yang Wei-T'e, of the Sung Dynasty, in the year 1054. He reported to the emperor: "I bow low. I have observed the apparition of a guest star. Its color was an iridescent yellow." The exploding star was visible in daylight for almost two years. It is shown in Figure 2.11 as it appears now.

With satellite-based telescopes that orbit the Earth far above the atmosphere, astronomers are able to obtain different and more penetrating views that reveal new features of the Crab Nebula not observed before. The top two panels of Figure 2.11 have about the same magnification but correspond to different wavelengths of radiation, namely optical and infrared. Our eyes are sensitive to the first and our skin to the second (heat). The infrared telescope detects hotter regions of the nebula, while an X-ray telescope detects regions where matter is in violent collision. The bottom two panels are clearer close-up views obtained by satellite-based optical and X-ray telescopes. These panels show the region that immediately surrounds the neutron star, the one that is the collapsed remains of the star whose explosion was recorded by Yang Wei-T'e in 1054. That very dense neutron star — only 15 miles in diameter — is spinning 30 times a second.

A neutron star acquires its rapid rotation in the same way that an ice-skater does — who, when whirling, draws in outstretched arms and spins faster. Similarly, when the iron core of the luminous star collapses to form a neutron star, it spins faster to conserve angular momentum. This is one of the laws of mechanics. Time lapse photography has revealed the expulsion of material from the location of the spinning neutron star at speeds near that of light. The X-ray photo reveals two jets, back to back, the one more visible than the other because of our perspective on the nebula.

The regular beat of concentrated radio pulses detected by huge antennas (Figure 2.12) is what first revealed, and still reveals, the existence of very dense neutron stars weighing as much as our Sun, but only about 20 kilometers across. Some of them rotate as many as hundreds of times a second. These are the powerhouses that can illuminate a whole nebula, such as the famous Crab Nebula, with a power equal to that of 100 000 suns. They are denser than the nuclei of atoms by a factor of 5 or more. Inside such stars there may exist matter in a form such as exists nowhere else in the universe today, but which was the form through which matter first passed at a time earlier than one hundred millionth of a second in the life of the universe. It is called *quark matter*.

Fig. 2.12. Arecibo radio telescope. The reflecting surface, or radio mirror, is 1000 feet in diameter and 167 feet deep, and covers, an area of about 20 acres. Being a radio telescope, it operates 24 hours a day. *Credit: NAIC — Arecibo Observatory, a facility of the NSF. Photo by David Parker/Science Photo Library.*

2.9   Milky Way

That pale luminous band of light across the sky that we call the Milky Way is as bright as 10 billion suns (Figure 2.13). Our Sun is one of the stars in the Milky Way galaxy. From our perspective on the Earth we have an inside view of part of our own galaxy. It consists altogether of about 400 billion stars, large and small, together with their planets as well as thousands of clusters of stars called *globular clusters* (Figure 2.14). Interspersed among the stars are vast clouds of dust, hydrogen, and helium (Figure 2.15). These clouds, if large enough, will incubate new stars.

The Milky Way is believed to look very much like the Andromeda galaxy viewed in Figure 1.12. It has a central bulge surrounded by a disk of spiral arms; both the bulge and the disk consist mostly of stars and a little gas

Fig. 2.13. Central portion of the Milky Way. The dust clouds are most visible, though almost all the mass is contained in stars. Light is absorbed by dust, which obscures the other side of the galaxy. *Credit: Axel Mellinger, University of Potsdam, Germany.*

and dust. Our Sun is about two thirds of the way out from the center of the galaxy, in the plane of the disk. We are looking toward the center rather than outward through the thin disk when we view the Milky Way, because that is the direction with the greatest concentration of stars and dust.

Light takes 90 000 years to cross the Milky Way. This huge disk is rotating; at the location of the Sun, the stars are moving at a speed of 250 kilometers per second about the galactic center. However, even at this tremendous speed, the Sun will have circled the center only 20 times, since it was formed about 4.5 billion years ago.

For every star in our galaxy having 10 Sun masses, there are about ten stars like our own; and for every star like our Sun there are ten or more with a mass of 1/10 Sun masses. Altogether there are about 400 billion stars of all masses. But the volume of the galaxy is so great that typical separations of stars are tens of millions of star diameters. Because of their vastness, galaxies can collide while doing little damage to the stars, though gravitationally distorting and perhaps occasionally disrupting the galaxies. Our galaxy weighs about 10^{12} suns but only about 1/10 of that is in the form of visible stars; most of it consists of an unidentified type of matter referred to as *dark matter*. Surrounding the central bulge of the galaxy out to the galaxy radius of 30 000 light years, there are the densely packed *globular clusters* of up to a million stars (Figure 2.14).

Fig. 2.14. The globular cluster 47 Tucanae, one of several hundred clusters that form a halo around the central bulge of the Milky Way. The density of stars in a globular cluster is up to a thousand times greater than that of the Milky Way. Clusters contain $10^4 - 10^6$ stars, which provides an estimate of the cluster mass in solar masses. Most globular clusters move in highly eccentric elliptical orbits that carry them on excursions across and far outside the Milky Way, to which they are bound by gravity. *Courtesy: David Malin with the 3.9-meter Anglo-Australian Telescope, and the Anglo-Australian Observatory.*

The mass and size of the Milky Way are fairly typical of other spiral galaxies. Galaxy masses fall within a factor of 100 of ours, and most spiral galaxies have radii similar to that of ours. The Milky Way and the Andromeda galaxy (Figure 1.12) are the principal members of a small group that is gravitationally bound to, and is on the periphery of, the Virgo cluster. The Virgo cluster is a group of about 2000 galaxies, which in turn is near the center of a *supercluster* of galaxies. The Milky Way is falling toward the Virgo cluster at a speed of 250–300 kilometers per second. Our galaxy is a typical one, yet contains a trillion suns, and around many of these suns are other planets, formed from parts of enormous molecular clouds as they collapsed and fragmented into galaxies of stars. It was heavenly fireworks, the like of which continues to light parts of the universe.

2.10   Universe Without a Center

The universe is about 15 billion years old. Because light has a finite, and not infinite, speed, the distance that it can travel since the Big Bang imposes a limit on how far we can see; we referred to the limit of the visible universe as *our* cosmic horizon. Let us imagine looking at a galaxy that is, say, 5 billion light years away. It has taken light 5 billion years to reach us. We

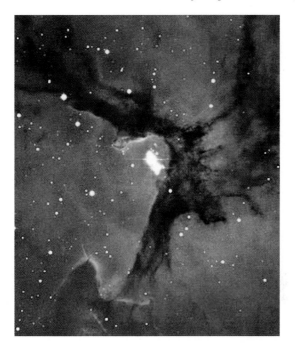

Fig. 2.15. Vast clouds of hydrogen mixed with tiny dust grains are distributed throughout the Milky Way. This cloud is a small part of the Trifid Nebula (M20) and is in the constellation of Sagittarius at a distance of about 3000 light years. Light and color are created by high energy gamma rays from star formation striking and ionizing hydrogen. Foreground stars are seen as bright white spots. *Credit: T. Boroson, AURA, NOAO, and N.S.F.*

see, now, in our time, how that galaxy actually looked 5 billion years ago. This time is referred to as the *look-back time.*

At Mount Wilson in southern California, Edwin Hubble (1889–1953) counted galaxies at ever-greater distances and found their number increasing in every angular patch of sky, just as if they were *evenly* distributed through space. His discovery has been confirmed with ever-more-powerful telescopes and sophisticated computer mapping devices no matter the search direction. The marvels of modern computers and large digital cameras on satellites reveal many details of our universe that stretch in time over billions of years and yet are recorded in a single photograph.

Figure 2.16 shows, by the use of color-coding, that at the earliest times the universe was uniform on the small scale of intergalactic distance, and at later times, uniform on ever-larger scales of clusters of hundreds of galaxies and then superclusters. This is what we expect; as time goes on, gravity

Fig. 2.16. This composite photograph of about 1/4 of the whole sky contains about 3 million galaxies. It is actually a *three*-dimensional view into the distant and therefore long-ago universe. Besides the obvious two spatial dimensions, the third is both a space and a time dimension —time because of the finite speed of light. The time coordinate is coded in color according to the apparent magnitude of the galaxies in each pixel. Fainter galaxies — which tend to be further away — are coded in red. We see *them* as they were about 10 billion years ago, and we note that they are distributed very evenly, which supports the conjecture of the general homogeneity of the universe. Galaxies of intermediate brightness are shown as green, and bright (close and young) galaxies are shown as blue. The younger galaxies that have formed clusters containing hundreds of galaxies are seen as small bright patches. Larger, elongated bright areas are superclusters and filaments. These surround darker voids where there are fewer galaxies. The time axis (color) thus shows progressively greater formation of structure as time passes. Even the structure is fairly uniform, as can be seen by comparing any one square inch of the photo with another. Small patches in the direction of close bright stars are obscured (the small cut-out patches). *Courtesy: S. Madox, W. Sutherland, G. Efstathiou, and J. Loveday. Facilities: UK Schmidt Telescope at Siding Spring, Australia, and Automatic Plate Measuring, Cambridge University.*

builds on small inhomogeneities to form larger structures, which are uniform on a scale that takes many structures into account. Thus, Hubble's conjecture is confirmed in great detail: the part of the universe that lies within our horizon is uniform on the *large scale*, especially at early times, and the same in every direction (isotropic). This discovery rivaled in importance Hubble's discovery of universal expansion because homogeneity and isotropy are exactly what is expected according to the interpretation of all the discoveries concerning our place in the universe, beginning with Copernicus. *We do not occupy a special place in the universe but rather a typical one.* This statement, recall, is known as the *Copernican cosmological principle.*

Imagine an inhabited planet in a distant galaxy: astronomers there could make discoveries and reach conclusions about their own planet and galaxy, and their place in the universe identical to those we have arrived at — that every point of view in the universe is the same, on the large scale, as any other — that all distant galaxies are rushing away from *them* in the same way Hubble found they are rushing away from *us*, and so on. If this were not so, it would imply that we were occupying the central position of the universe and this would defy a sense of perspective. It being accepted, therefore, that we occupy a typical position, let us imagine going back in history — back in time. In the distant past, toward the beginning of time when all matter began its rush into the future, there is a time so early that light will not have had sufficient time to travel from that other galaxy to ours. Just as the cosmic horizon for any observer anywhere in the universe is growing as time passes, it is ever-smaller as one goes back in time toward the beginning. It becomes smaller and smaller in the past; every part of the visible universe was beyond the horizon of every other part near the beginning of cosmic time.

We can push this line of reasoning back to the time when the first nucleons, electrons, neutrinos, and light emerged from the great inferno called the Big Bang — a time so early that the horizon of every particle in the universe did not extend beyond itself.

Where is the center of the universe? Every particle *then* was rushing away from every other particle, just as every galaxy *now* is rushing away from every other galaxy. *Clearly, there is no center.*

2.11 Cosmic Horizon as Small as a Nucleon

How long after the beginning was that moment in time when the cosmic horizon for every nucleon was no larger than the nucleon itself? Knowing the velocity of light and the size of a nucleon, the answer can be found immediately. It is the size of a nucleon divided by the speed of light. This turns out to be about 3×10^{-24} seconds ($= 0.000000000000000000000003$ seconds).[15]

---

[15] Have I counted the zeros correctly?

We can think back to this incredibly short time after the beginning, using laws of physics as we know them now and especially Einstein's general theory of relativity. Some cosmologists consider even earlier times, back to what is called the Planck time, earlier than which relativity and the other laws of nature such as hold in our universe now could not have held then in a universe so small. Unless the relativity and quantum theories can be merged into one overarching theory, we can never learn what happened during the Planck era — we do not even know if time had a meaning.

2.12   Box 4

---

4   Hubble Constant and Universe Age

The measurement of the Hubble constant is difficult, and various means give somewhat different answers. It is an ongoing effort to accurately determine it. Hubble estimated it by plotting the distance of nearby galaxies as determined by parallax — good to 100 ly ($= 3 \times 10^{15}$ km) — versus the velocity of recession as determined by the redshift or Doppler shift. Extrapolation to larger distance is not reliable and other methods have to be used. A current estimate of the Hubble constant is

$$H_0 = 200 \; \frac{\text{km}}{\text{s}} \; \text{per} \; 10^7 \; \text{ly} \,.$$

This yields, for the universe age estimated as the inverse of Hubble, $1/H_0 = 15 \times 10^9$ years.

---

3 Big Bang

Some say the world will end in fire,
Some say ice.
From what I've tasted of desire
I hold with those who favor fire.
But if it had to perish twice,
I think I know enough of hate
To say that for destruction ice
Is also great
And would suffice.

— Robert Frost, *Fire and Ice*

3.1 Hubble's Discovery

Not only is the universe stranger than we imagine, it is stranger than we can imagine.

— Sir Arthur Eddington

From the time of Copernicus, we have known that the Earth is not at the center of the universe, nor even the Sun. Rather, the Sun is an insignificant star, one of an enormous number in a starry heaven. Then Herschel discovered that the starry heaven we see with our naked eyes is but one of many that he called island universes. Now we know that there are billions of these island universes, which we now call galaxies.[1]

With the powerful telescope at Mount Wilson in southern California, Edwin Hubble (1889–1953) counted galaxies at ever-greater distances and found their number increasing in every angular patch of sky, just as if they were evenly distributed through space. This has been confirmed time and again in more recent deep surveys, no matter the direction. Evidently, the part of the universe that lies within our horizon is *homogeneous* on

---

[1] There are an estimated 10 billion galaxies in the *visible* universe.

the large scale, and *isotropic*. But, surely we are not at the center of the universe. Rather, our position is typical unless we make the preposterous assumption that we, on this small planet that orbits an insignificant sun in a galaxy containing billions of suns, are at the center. Consequently, another distant observer, but near the edge of our horizon, would observe the same isotropy and homogeneity. And so on for a third observer, lying within his horizon, but far beyond our own. And, like Hubble's other great discovery, each observer would see all distant galaxies receding at a speed in direct proportion to the distance from *him*.

To visualize an expansion with the above properties, consider a one-dimensional universe. Imagine a very long elastic string. Lay it out and attach tiny buttons spaced one inch apart to represent galaxies. Stretch the elastic so that the distance doubles between every two *adjacent* buttons. If, from any button, one looks down the line two buttons, its distance from the first, which was two inches before the stretching, is now double that. And so on; at the location of any button, all other buttons in either direction will have doubled their distance, and all in the *same* amount of time. Therefore the speed of each button away from the chosen one is proportional to its distance. And this is true no matter which button we wish to measure from. We have here a simple example of how uniformity implies Hubble's law for the velocity of recession — meaning, as it does in this example, that the view from any button up and down the line is the same view as from any other button. The converse is also true: If the velocity of recession is found to be proportional to distance from the observer, then the universe must be uniform, or, as we frequently say, homogeneous.

The appellation "Big Bang" was used first by Fred Hoyle as a humorous way of deriding the cosmology that now bears that name. Bondi, Gold, Hoyle, and Narlickar argued for a steady state cosmology rather than an evolution from a hot dense beginning. Most cosmologists are convinced that the existing evidence along the lines discussed in the next section points to a hot beginning. However, in some sense, Hoyle's derision plays a trick on those of us who believe the Big Bang theory of cosmic evolution. The universe was neither big, nor was there a bang. "Bang" conjures up an explosion in the imagination, and an explosion occurs from a point, or at least a definite region out of which the gases are driven into the exterior space by a steep pressure gradient between the two regions. The universe has no center and no edge, as we learned in Section 2.10, page 54. Thus, there can be no pressure gradient. As George McVittie wrote, "... it is unfortunate that the term 'big bang', so casually introduced by

Hoyle, has acquired the vogue which it has achieved." Perhaps no one has found an apt name for that pregnant moment of creation because *nothing* in everyday experience resembles it. At any rate, we seem to have no other than this ungracious name for that moment of ineffable glory when the universe began its ongoing journey, *creating time and space as it expands.*

3.2 Evidence for a Big Bang

> In science one tries to tell people, in such a way as to be understood by everyone, something that no one ever knew before. But in poetry, it's the exact opposite.
>
> — P.A.M. Dirac

The term "Big Bang" has been floating in the air for a few years. Why are almost all astronomers and cosmologists persuaded that we can trace the universe back to an instant in time when all the matter of the visible galaxies was enormously hot and dense?[2] The question goes to the very heart of nature. By what marvelous processes did all we see *on* the Earth, all we see *from* the Earth, come to be? What we have to go on are the laws of nature as they have been uncovered over the centuries and the application of those laws to the evidence that can be obtained by looking at the universe in our neighborhood, and looking backward in time by studying its most distant reaches.

There are four reasons for believing that the universe started life with a "bang". The first observations to suggest a small, hot, and dense beginning were the discoveries of Hubble — that the universe is expanding, and that the universe is uniform and the same in all directions. We have seen what important conclusions can be drawn from these discoveries, summarized by the cosmological principle.

The next observation that points to a very hot beginning was the discovery by Penzias and Wilson of the *cosmic background radiation*. The universe is *pervaded* by radio waves whose temperature is 3 degrees Kelvin, close to absolute zero on that scale.[3] On the more familiar scale, radiation

---

[2]P.J.E. Peebles, D.N. Schramm, M.E. Turner, and R.G. Kron, "The case for the relativistic hot big bang cosmology", *Nature*, Vol. 352 (1991), p. 769.

[3]At 0 degrees Kelvin there is *no* thermal agitation of molecules, atoms, or anything; there is absolute stillness aside from a quantum twitching — a manifestation of the uncertainty principle. Since the early measurements, the background radiation temperature has been known more accurately as 2.7277 K.

in the universe is at -270 degrees Centigrade. The uniformity of this radiation from all parts of the sky tells us that it was not caused by any of the objects that exist in the universe now, like stars and galaxies, for they are hot *concentrated* sources. Evidently, the uniform low temperature radiation is the faint glow from the time of a very hot and uniform universe, now very much cooled by the expansion. That story is the subject of the next section.

The other two pieces of evidence have to do with what happened in the universe during its expansion — the synthesis of the light elements like deuterium and helium in the first few minutes, and the formation much later of galaxies and clusters of galaxies. As we will see shortly, the evolution of the universe can be divided, as is often the case in physics, into two parts according to their scale: one is the large scale expansion of the universe and the other is the growth of structure within the expanding universe. We are dealing with the first part in this chapter; we leave the last two pieces of evidence of a Big Bang until later chapters.

The enormity of space and time spanned by the present universe provides us with a great advantage in fathoming its life history. When we look at the nearby parts of the universe, like our own and neighboring galaxies, we see what the laws of nature have wrought from the *initial* conditions of the universe 15 billion years ago. Using powerful telescopes, the astronomer, when looking out at greater distance, is also looking back in time. The distance is not so valuable in itself as the fact that the light that arrives at his lens started its journey long ago. The telescope in this sense is a time machine. It does not transport *us* back in time, but shows us what the universe was like back then. And by looking ever deeper into space, the astronomer is looking ever farther back in time. It is like running a movie backward.

That is very valuable but it is not yet the full story. We see what the laws of nature have produced, step by step from the preceding frame, but the scientist seeks more than a factual history. He asks how and why. The historian does too, but that is not always so evident when we "learn" history as an account of the past. When the scientist can provide all the hows and whys linking one step to the other, we can have some degree of confidence in what he has to say about beginnings. Many of these links have been made, as this book recounts. However, it is possible that one "how and why" will remain forever a mystery. How and why did this amazing universe begin? But from shortly *after the beginning* we have the tools to answer the hows and whys, though not in all cases do we yet have the necessary data.

Let us look now at the spectacular evidence that the universe was once enormously dense, hot, and uniform. The evidence comes from an early time when the universe was only about 300 000 years old and the temperature was about 3000° (see Section 5.4.6). Before that time and at the corresponding higher temperatures, radiation and matter interacted strongly. But at about that time, radiation decoupled from matter and streamed freely through the universe. As the universe expanded, the wavelength of radiation expanded with it; the temperature decreased inversely to the expansion scale. This untouched radiation from that early time of 300 000 years is the *cosmic background radiation* discovered by Penzias and Wilson; it is now a faint glow from the past.[4]

3.3   A Day Without Yesterday

The young Belgian theological student Georges Lemaître (1894–1966), later a priest, struggled to reconcile the biblical account of creation with Einstein's theory of gravity (general relativity). He, and independently the Russian mathematician Friedmann, understood the new and revolutionary implications of Einstein's theory for expansion of the universe. To further his task, he took up the study of astrophysics and cosmology at Cambridge University and the Massachusetts Institute of Technology, where he was awarded his doctoral degree in 1927.[5]

Lemaître published his early ideas on the birth of the universe in an obscure Belgian journal, where it passed unnoticed by the principals in the field.[6] In that work, he applied Einstein's theory of gravity to cosmic expansion and conceived the notion of the "primeval atom", which is the term he used to describe the universe at its beginning. His prescient notion of cosmic acceleration, which he included among the possible universes, has actually been confirmed only in the past several years and ranks among the great cosmological discoveries of all time (Section 7.3).

It is all the more remarkable that Lemaître's publication of his audacious theory of the beginning of the world actually preceded Hubble's 1929 discovery of universal expansion. After he had learned of Hubble's discovery, he quoted that work in his subsequent publications to support his ideas.

---

[4]In the scientific literature the background radiation is usually referred to as the "cosmic microwave background radiation", abbreviated as CMBR.

[5]G. Lemaître, *Nature*, Vol. 127 (1931), p. 706.

[6]*Annales de la Societe Scientifique de Bruxelles*, 1927.

Meanwhile, Hubble, in California, was unaware of the young priest's theory, which had foreshadowed his own great discoveries.

In 1933, both Lemaître and Einstein gave a series of lectures in California. It has been reported that after the Belgian priest detailed his theory, Einstein stood to applaud, saying: "This is the most beautiful and satisfactory explanation of creation to which I have ever listened." Several years later, Pope Pius XI inducted Lemaître into the Pontifical Academy of Science.

Fig. 3.1. The Belgian priest G. Lemaître (1894–1966), who studied cosmology besides theology. He used Einstein's theory of general relativity to study model universes, and was the first to realize the possibility that creation began with a "Big Bang". Fred Hoyle was the one who coined that name. Evidence that that is indeed how the universe began was serendipitously discovered by Penzias and Wilson at the Bell Laboratory.

Lemaître's beautiful theory of creation and those of other early cosmologists, including Einstein, are described in the next section. Their ideas play an essential role in the interpretation of the faint glow — the cosmic background radiation — seen by sensitive radio antennae in the present time that is a relic from the hot past at a time of only 300 000 years after the beginning.

George Gamow (1904–1968) and his collaborators, Ralph Alpher and Robert Herman, suggested in 1948 that a relic from an early era might pervade the universe here and now.[7] Gamow, born in Russia, later a US citizen, developed the theory of alpha radioactivity—the spontaneous emission of alpha particles (helium-4 nuclei) from certain atoms. At that time Gamow was a young man in the institute of Niels Bohr in Copenhagen.

Gamow realized from his early work on apha decay that very high temperatures would be needed to forge the elements of the periodic table from neutrons and protons; the process is referred to as thermonuclear fusion. He proposed that the elements were forged in the early universe when it was very hot. Gamow was only partially correct. Only very *light* elements, mostly helium, were produced in the early universe; the universe passed through the favorable temperature range much too quickly to synthesize the long chain of reactions that would produce the heavier elements. The intermediate mass elements from carbon to iron are forged from hydrogen and helium during the long lifetime of stars, both those we see now, and those that have long ago died. The energy produced by these thermonuclear reactions is what we see as light and feel as heat from the Sun and other stars. The *heavy* elements are made in small abundance in the fiery material expelled by a star in a supernova explosion at the end of its life.

Another extremely important idea emerged from Gamow's work. As the universe expanded it cooled, but radiation that was characteristic of an earlier time — about 300 000 years after the light elements were forged — would remain in the universe as a messenger from that early epoch.

Why the temperature stamp of the early universe, modified in a predictable way only by the expansion, would survive for 15 billion years is another story, which we take up in Section 5.4.6, page 144. All that the cosmic expansion did to radiation *since* that long-ago time was to reduce its temperature in a very precise way: the temperature would decline inversely to the factor by which the expansion had increased ($T \sim 1/R$). This is the effect of the Doppler shift about which we commented earlier (Section 1.3, page 22). Gamow, with Alpher and Hermann, calculated the present temperature to be about 5 Kelvin, close to absolute zero, at which temperature there is no heat at all. Considering the uncertainties involved, which had to do as much with plasma physics as with cosmic evolution, this turned out to be amazingly accurate.

Arno Penzias and Robert Wilson, working at the Bell Laboratory in New Jersey, knew nothing of this work when they set about making measurements of radio noise in the sky; their aim was to improve satellite

---

[7]George Gamow, *Creation of the Universe* (The Viking Press, second printing, 1959).

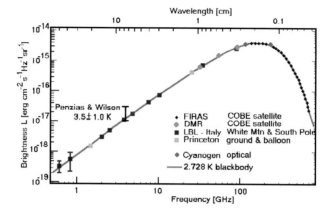

Fig. 3.2. Spectrum of the cosmic microwave background. The various observations lie on the violet curve, which represents the radiation from a perfect blackbody with a temperature of 2.728 Kelvin. The blue bar represents the original 1965 measurement by Penzias and Wilson. The green dots represent the 1941 measurement by McKellar. The significance of that measurement was not realized until after the measurement of Penzias and Wilson. We encounter later the significance of the shape of the curve that is so accurately determined by the measurements. *Credit: G. Smoot, LBNL, Berkeley, and the COBE satellite project, NASA. Color added by D. McCray.*

communications. But a persistent annoying hiss was detected by their radio antenna, which for a long time they thought was interference, possibly in the antenna itself or the radio amplifiers. Eventually it became evident that the radio signal that they were detecting came from all directions of the sky. It is in fact the same radio signal that contributes a small fraction of the snowlike interference on television sets. They characterized the noise by a temperature in a way we will understand later when we discuss in detail the material and radiation content of the early universe before there were galaxies. Penzias and Wilson, who won a Nobel Prize for their discovery, had no idea what this radio message signified until they were put in touch with Robert Dicke and Jim Peebles at Princeton University.

These two scientists, Dicke the experimenter and Peebles a young theoretician recently arrived from Canada, were in fact preparing a specialized radio antenna for the express purpose of searching for light that had come to us across the millennia from a time when the cosmos was young. Like Penzias and Wilson, Dicke and Peebles were unaware of George Gamow's pioneering work, but they had reached a conclusion similar to his. When Dicke received the phone call from the Bell Lab team telling him that a constant 3 degrees Kelvin signal had been detected from intergalactic space at the Holmdel antenna, Dicke murmured to his collaborators: "We've been scooped." What had been measured was a cold relic from the once-hot past. The two groups agreed to publish the findings and the interpretation as companion papers in the *Astrophysical Journal* in 1965.[8]

The discovery of the relic radiation, once-intense gamma and X-rays, now cooled to very low energy photons in the radio, infrared, and optical bands, was one of the three most important discoveries in cosmology made in the 20th century. They are the universal expansion discovered by Hubble, the cosmic background radiation by Penzias and Wilson, and, if it holds up under the intense scrutiny it is being subjected to, the *accelerating* expansion of the universe by S. Perlmutter and G. Goldhaber.

The 1965 discovery of Penzias and Wilson was a turning point in cosmology. Thereafter it developed into a science rather than an art. When one looks at the single data point they measured on the graph in Figure 3.2, one might ask what all the fuss was about. But all of the more accurate data obtained since then should not obscure the fact that their discovery was strong evidence for a very hot beginning and provided the rationale and stimulus for the marvelous discoveries made since then. The more accurate data displayed in the graph, which was much later detected by cryogenically cooled detectors carried aloft in balloons and satellites, confirmed the interpretation beyond doubt, and added a new dimension, not visible in this graph.

The cosmic radiation, now a mere three degrees above absolute zero, came, untouched, from an early time when the universe was more than a thousand times smaller and very hot, about 3000 Kelvin. Its present lower temperature is merely the Doppler effect of the cosmic expansion. Later experiments found that the cosmic radiation had a uniform temperature to very high degree so that conditions at that early time were indeed very uniform. If they had not been, there would be little chance that the history of the universe could be traced. As we will see in the next section, neither Einstein's theory nor any other would have enabled the cosmologist to trace the history in a quantifiable and therefore verifiable, or falsifiable, way.

The high degree of uniformity of the background radiation confirmed that the approach of early cosmologists like Einstein, Friedmann and Lemaître made sense. However, as important as that was, the extreme accuracy of the observations made with equipment on the COBE satellite by

---

[8]A.A. Penzias and R.W. Wilson, *Astrophysical Journal*, Vol. 142 (1965), p. 419; R.H. Dicke, P.J.E. Peebles, P.G. Roll, and D.T. Wilkinson, *Astrophysical Journal*, Vol. 142 (1965), p. 414.

Fig. 3.3. In April 1992, George Smoot announced that the long-sought evidence for the seeds on which gravity worked to grow galaxies, clusters of galaxies, and clusters of clusters had been found by the team that he led (COBE DMR team). he seeds had been found in minute temperature variations in the early universe and were registered in instruments carried on NASA's Cosmic Background Explorer satellite.

the team led by George Smoot (see Figure. 3.3) discovered the long-sought variations in the early universe. NASA's COBE (Cosmic Background Explorer) satellite mapped the intensity of the radiation from the early Big Bang and found variations so small they had to be the seeds on which gravity worked to grow the galaxies, clusters of galaxies, and clusters of clusters that are observed in the universe today. Not merely the existence of a slight clumpiness but the quantitative degree of *clumpiness*—the very seeds—needed to form galaxies is being tested today (see Figure 3.4).

3.4   Temperature Measured at an Earlier Time

The high degree of uniformity of the cosmic background radiation coming from all directions of the sky is persuasive evidence that it pervades the entire universe and is a relic from a time long before there were galaxies and stars. Other sources of radiation like galaxies and stars are not uniform on the sky, but are concentrated sources of radiation. As if any further proof of the early origin were needed, the temperature of the cosmic background radiation *has* been measured *as it was* at an earlier epoch, when the universe was little more than 1/10 its present age, was about 1/3 as big as it is now, and its temperature was 3 times hotter.

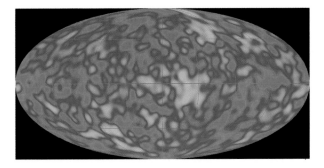

Fig. 3.4. This is a map of the sky showing slight variations by color in the temperature of the cosmic background radiation made by the COBE DMR team led by George Smoot. The variations are extremely slight, only a few parts in 100 000, and represent slight variations in the density of matter in the early universe that are thought to be the seeds of later galaxies. *Credit: COBE satellite, NASA.*

How can this be done? How can measurements by us in our time be made on the early universe? Recall that when the astronomer looks deep into space, he is looking back in time. With powerful telescopes the temperature in distant molecular clouds has been measured; the measurements are difficult so that an exact number has not been determined.[9] Rather, the temperature was found to lie in the range $6.0 < T < 14$. This is in satisfactory agreement with the expected temperature of 3×2.74 Kelvin.[10] These measurements are a beautiful additional confirmation of the cosmic origin of the background radiation and of the way its temperature depends on the scale of the universe.[11]

3.5   Model Universes

The most incomprehensible thing about the world is that it is comprehensible.

— Albert Einstein

---

[9]R. Srianand, P. Petitjean, and C. Ledoux, *Nature*, Vol. 408 (2000), p. 931.

[10]The temperature of the background radiation would have been higher than it is now (2.74 Kelvin) by the factor by which the universe has expanded since then, namely R_0/R, and it would have been younger by the square of that factor, namely $(R_0/R)^2$. In terms of redshift, those factors are $1/(1+z)$ and $1/(1+z)^2$, respectively. The relation between expansion factor and time depends on the cosmological model and the values of the parameters that enter the model.

[11]The details of this dependence will be given on page 73 and in boxes starting on page 153.

The cause for the birth of the universe will remain forever a mystery, beyond the power of science to reveal. However, in a broad sense, there are surprisingly few alternative courses the universe could have followed once it was born. The Belgian priest Georges Lemaître discovered that Einstein's general theory of relativity is *the* theory that is able to delineate the possibilities. He actually displayed by calculations and diagrams the different possible scenarios — a universe that expanded and then recontracted, a universe that expanded forever but in a coasting manner, and a universe that broke into an accelerating expansion, as has recently been discovered to be the actual course of *our* universe.

Indeed, Einstein himself understood the scope of his theory. But the universe appeared to be static at the time he first thought of using his theory in this way. He therefore introduced an extra term, a constant, into his gravitational equations — which did not need to be there but which was allowed — for the express purpose of maintaining a static model of the universe (Box 5). That term is called the "cosmological term", and its constant value is denoted by Λ. It is referred to as *cosmological* because it is too weak to affect the structure of individual stars but being a *constant* density its effect is cumulative. Therefore, the cosmological term can act on the entire cosmos.[12] About two years later, Hubble, using the telescope at Mount Wilson in southern California, discovered the universal expansion and the law that bears his name. Recent discoveries have proven that indeed the cosmological term, once deemed unnecessary, is actually real and represents the mysterious *dark energy* that drives the universe toward an exponential expansion. Evidence of the existence of dark energy has been found only in the past several years (see Section 7.3).

Einstein's theory of gravitation has been verified in a number of ways. Einstein himself pointed out one of the immediate triumphs of the theory. Among the tests that the theory passed immediately after its publication in 1916 was an explanation for the precession of the axis of Mercury's elliptical orbit around the Sun. The orbit follows an elliptical course according to Newton's gravity and Kepler's discovery of the laws of planetary motion. But the axis of the ellipse also rotates around the Sun, a striking feature that Newton's gravity could not account for (except for a rotation caused by the gravity of the other planets which could be calculated but did not

---

[12]General relativity can describe single stars without the cosmological term, and a correct description of them restricts the size of the cosmological term to be very small in the relevant units.

make up the full amount of the rotation). Einstein, himself, performed this calculation, which agreed very closely with the measurements.

The same effect, but much more dramatic, has been observed in the orbit of the pair of Hulse–Taylor neutron stars. Each of them has a mass that is a little larger than that of the Sun. However, the radius of the orbit is much smaller than that of Mercury. So the effect is 3500 times larger for the neutron stars. The radiation of gravitational waves that are ripples in spacetime removes energy from the orbit of the pair of neutron stars, causing the orbit to shrink very slightly as time goes on. The extent of this effect has been measured over a period of more than 20 years and provides a test of Einstein's theory to better than one percent.[13] In about a hundred million years the two stars will collide, sending out a distinct pattern of gravitational waves.

Of course, we cannot expect that there will be human observers to witness the effect of the collision of the Hulse–Taylor neutron stars. However, the universe is vast and we now have the means of seeing to a great distance. Therefore, in our own time, in the whole Milky Way galaxy, in the 2000 galaxies of the nearby Virgo cluster, and in all the myriad not-too-distant other galaxies, there must be other such binary systems as the Hulse–Taylor binary neutron stars. Some of these binaries are expected to produce the characteristic inspiraling signal that will radiate outward as ripples on spacetime. These, among other gigantic cosmic events, are the objects of the search by the gravitational wave detectors that have been and are under construction in various parts of the world.

The most ambitious program, called LISA, is sponsored jointly by the European Space Agency (ESA) and the National Aeronautics and Space and Administration (NASA)(see Figure 3.5). This mission will involve three spacecraft flying approximately *5 million* kilometres apart in an equilateral triangle formation. Together, they act as a Michelson interferometer to measure the distortion of space caused by passing gravitational waves. Lasers in each spacecraft will be used to measure minute changes in the separation distances of free-floating masses within each spacecraft.

General relativity is a theory that describes the evolution of the universe on the *large scale*, such as how the average temperature and the density of matter were diluted by the expansion. It does not describe the details, such as how large clouds of diffuse matter in the early universe began to aggregate still more matter and then began a gravitational collapse, fragmenting as

---

[13]C.M. Will, *Was Einstein Right?* (Oxford University Press, 1995).

Fig. 3.5. An artist's rendition of a future mission to detect gravitational waves. Three spacecraft will fly *5 million* kilometers apart in triangular formation. Together, they act as a Michelson interferometer to measure the distortion of space caused by passing gravitational waves. A distant disturbance of spacetime consisting of a black hole merger is shown near the top of the figure. *Credit: European Space Agency.*

they did so into galaxy clusters, galaxies, stars, and finally planets. These are separate problems of fine detail that take place against the background of evolving density and temperature.

That these structures are really a fine structure can be understood, given that the diameter of the visible universe is 200 000 times greater than that of our own galaxy, and even many times larger than the diameter of a typical galaxy. The galaxies in a patch of sky viewed with powerful telescopes appears almost as dust grains spread more or less uniformly, as in Figure 2.16, especially as regards those that formed early. So the evolution of the universe on the large scale, and the evolution of structures within it, such as planets, stars, galaxies, and clusters of galaxies, can be separated into two problems.

The *uniformity* of the universe in all *directions* is referred to often as the *homogeneity and isotropy* of the universe. These are key observations that make it possible to describe the evolution of the universe in a quantifiable way, and have been referred to many times previously in this book. Recent experiments carried aloft in balloons and satellites that measure a whisper from the long-ago past, a radio signal called the cosmic background radiation, first discovered by Penzias and Wilson, have confirmed

Fig. 3.6. Einstein sailing near Princeton, 20 years after the publication of his most famous paper, the one on general relativity.

the isotropy and homogeneity of the early universe to a very fine degree. The slight departure from homogeneity that is observed is about the degree that would be required to account for the condensation of great clouds in which galaxy clusters and galaxies started to form about 12 billion years ago.

The above observations place a requirement on the geometry of space. It can have but a single curvature at any given cosmic time. However, the curvature need not remain constant in time; it can everywhere vary in time in the same way. Let us see how the past and future of the universe might unfold according to the very limited number of models allowed by general relativity, cosmic isotropy and homogeneity.

3.6   Scale Factor of the Universe

> Nature is an infinite sphere, whose centre is everywhere and whose circumference is nowhere.
>
> — Blaise Pascal, *Pensées* (1660)

At this stage in our study of cosmology, we are considering the universal expansion on the large scale, not the details of structure formation such as galaxies. For the purpose of describing the expansion, we can think of the universe as a cosmic dust of particles, which may be nucleons or galaxies, depending on the time. Under the assumption of universal homogeneity, and isotropy (cosmological principle), the expansion must appear the same to all observers anywhere at any particular cosmic time.

To give quantitative meaning to such an expanding universe, we may therefore introduce a *scale factor* (sometimes called an expansion factor) that is growing with the expansion of the universe. We denote it by $R(t)$. It depends only on time; not on position. It is everywhere the same because of the assumed universal homogeneity and isotropy of the universe. Thus the distance d_0 between any two galaxies at time t_0 and the distance d between them at time t is given by

$$d = d_0 \left[R(t)/R(t_0) \right] .$$

If $R(t)$ is large in the sense that a sphere of such a radius contains a vast number of marker particles, be they nucleons or galaxies, depending on the stage of expansion under consideration, then, according to the cosmological principle, all such spheres of the same radius are equivalent.

Einstein's theory of general relativity, which we can also refer to as the theory of gravity, provides the means by which we can learn how the universe expands as time passes. His equations are many in number but for a universe that is the same in all *directions* as viewed from any *location*,[14] they reduce to a single dynamical equation which describes how the universal scale factor $R(t)$ develops with time.

That single equation is called the Friedmann–Lemaître equation and it is

$$\dot{R}^2(t) = \frac{1}{3} \left[8\pi G \, \rho(t) + \Lambda \right] R^2(t) - k \, c^2 .$$

The dot over the scale factor R denotes the rate at which R changes with time $(\dot{R} = dR/dt)$; in other words, $\dot{R}(t)$ is the relative velocity of the expansion at time t (see Box 8). Cosmic time is denoted by t, past, present, or future, measured from the beginning of the expansion; G is Newton's gravitational constant; $\rho(t)$ is the average mass density per unit volume (which includes the mass equivalent of radiation) at time t. This equation was first derived from general relativity by the Russian mathematician Alexander Friedmann in 1922, and independently by the Belgian priest and cosmologist Georges Lemaître in 1927, who explored its solutions for an *acceleration* in the expansion of the universe.

---

[14]Isotropic and homogeneous as assumed by the cosmological principle.

Looking backward in time, each R shrinks but to different centers. Where then is the center of the universe? Clearly, as we have already reasoned (Section 2.10, page 54), there is no center. *At any cosmic time, the universe is expanding in the same way from every location.*

Of the two remaining quantities in the Friedmann–Lemaître equation, Λ denotes what is called the *cosmological constant*. Albert Einstein introduced this term into his gravitational equations because their form permitted such a generalization, *and* at that time it was thought that the universe was static rather than expanding: a small positive value of Λ prevented his particular model of the universe from *collapsing* under its own gravity, which is why Einstein introduced the term in the first place. However, Hubble discovered shortly afterward that the universe is actually expanding. In terms of the scale factor and its time derivative, the Hubble parameter is given according to its meaning by

$$H(t) = \dot{R}/R \,.$$

Einstein could have predicted the expanding universe from his equations before Hubble's discovery had he not been restrained by the general opinion of his time that the universe was static. This was one of the few occasions when Einstein allowed himself to be deterred by current opinion. More frequently, he struck out on his own, making mistakes sometimes, but always with a boldness of thought that made his era the most fruitful in the physical understanding of the world. He established Planck's constant, h, and the discrete energies of oscillating particles producing radiation, as an *actual* quantization of radiation itself, thereby making a great leap toward quantum theory.[15] Einstein rejected the Newtonian concept of space and time, and showed Newton's laws of mechanics to be a special case of his own theory, valid only for small speeds (compared to c) and places of weak gravity. And over a period of some 12 years, he established special relativity as the framework for mechanics when speeds are closer to that of light, and general relativity under all circumstances in the universe, excepting the extreme conditions at a time from the beginning that is almost infinitely small compared to a second.

The recent discovery that the cosmic expansion seems actually to be accelerating would have been even more astonishing in Einstein's day than in ours. If true, this discovery suggests the presence in the universe of a form of uniformly distributed energy of an unknown nature, which therefore is called *dark energy*. The cosmological constant, which rested so long in the

---

[15]Planck himself was somewhat vague about the meaning of h.

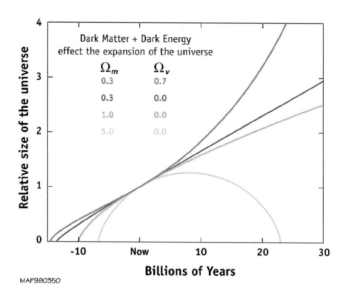

MAP990350

Fig. 3.7. The size of the universe, in arbitrary units, as a function of time for four hypothetical universes. They all start with a bang, but at different times in the past. One recollapses, two coast forever, and the top curve corresponds to accelerating expansion. The top curve corresponds to the actual measured values of the matter and vacuum energy (also known as *dark energy*) of the universe. The top curve represents *our* universe; the bottom one, "the big crunch".

dustbin of physics ideas, now occupies a central place in the evaluation of the parameters that govern the evolution and ultimate fate of the universe. Of these things we will want to say much more at a later point in the book.

The Friedmann–Lemaître equation depends on one remaining quantity. The homogeneity and isotropy observed in the universe permit *only* a universal spatial curvature, k (Box 7). There are three possibilities which can be represented by a positive, zero, or negative value of k. They correspond to a space curvature as listed in the table and as illustrated in Figure 3.8.[16]

$$
\begin{array}{ll}
k = +1: & \text{positive curvature (as for a sphere)} \\
k = 0: & \text{flat spatial curvature (as for a plane)} \\
k = -1: & \text{negative curvature (as for a hyperboloid)}
\end{array}
$$

---

[16]The origin of the parameter k lies in a freedom in the choice of the metric for a uniform homogeneous universe.

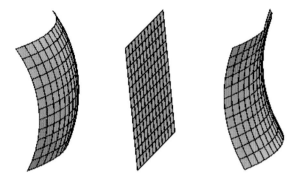

Fig. 3.8. The three possible spatial geometries of the universe at a slice of cosmic time. They correspond to $k = 1, 0, -1$ from left to right.

The long term behavior for each such hypothetical universe is different from the others, being closed or open, and accelerating or nonaccelerating expansion, depending on the value and sign of the cosmological constant. Let us see what the future of the universe might be, and what the present observational evidence suggests.

3.7   Theories of Expansion

When we learn by solving the theory of how the scale factor, R, evolves with time, we will easily deduce much more about the large scale properties of the universe as it developed after the beginning.[17] It will be possible to trace also how the temperature, the density of matter, and the density of radiation varied with time during the universal expansion. And when these things are known, a direct link will have been forged between the large scale smooth behavior as described by R and the formation of structure such as the elements of the periodic table, the galaxies, the galaxy clusters, and the stars out of which they are made. The whole picture, from near the

---

[17]The mathematical theory based on classical general relativity has a singularity at $t = 0$. However, there is very good reason to believe that the classical theory should not be pressed that far. At very short distance, relativity and quantum theory need to be melded into a single consistent theory of *quantum gravity*. That synthesis has not been achieved. We must be content in the meantime to remove from consideration times earlier than what is called the Planck time (page 121). This is an incredibly small 10^{-43} seconds, so as time is measured, we do not miss much. On the other hand, the greatest secrets of the cosmos are hidden behind that barrier.

beginning of time till now, must fit into an interlocking pattern; any persistent dissonance will betray a weakness in some of the original assumptions. These, in turn, will require revision until a perfect whole is found.

For, of course, the outcome of such calculations will depend on assumptions that are made concerning broad properties of the universe at the beginning, such as the general nature of the geometry of the universe, and whether there is sufficient matter in it that gravity acting on that matter will bring the expansion to a close someday. We will describe these universal "parameters" in detail in a later chapter. What is most important is that they are in principle subject to measurement. More than that — cosmologists already have some fairly convincing evidence as to what their actual values are.

3.7.1   *The big crunch*

> Don't worry if your theory doesn't agree with the observations, because they are probably wrong. But if your theory disagrees with the Second Law of Thermodynamics, it is in bad trouble.
>
> — Sir Arthur Eddington

This model universe was first discussed independently by a Russian meteorologist, Alexander Friedmann, and a Belgian priest, Georges Lemaître. They independently realized that one of the possible fates of a universe was to recontract to the very hot and dense condition it had started in after a long period of expansion. This model universe is often referred to by the irreverent phrase "the big crunch". Friedmann submitted his paper, with the surprising conclusion that the universe may be expanding, to the German physics journal *Zeitschrift für Physik*, whose editor sought Einstein's advice as a referee. He wrote back saying: "The results concerning the nonstationary world, contained in [Friedmann's] work, appear to me suspicious. In reality it turns out that the solution given in it does not satisfy the field equations [of general relativity]."

However, Friedmann was confident of the results that he had obtained from Einstein's theory. He wrote to Einstein, beginning: "Considering that the possible existence of a nonstationary world has a certain interest, I will allow myself to present to you here the calculations I have made...for verification and critical assessment...." Meanwhile, Einstein had already left for Kyoto and did not return to Europe for several months. Then, by chance a friend of Friedmann's met Einstein at Ehrenfest's house in Leiden and described his colleague's work; Einstein saw his error and immediately wrote to the journal's editor: "... my criticism [of Friedmann's paper]... was based on an error in my calculations. I consider that Mr Friedmann's results are correct and shed new light." Altogether, the publication of Friedmann's paper took little more than a year from the time it was submitted to the journal. Compared to modern times, even though editors, authors, and referees now make routine use of e-mail, this delay was not extraordinary.

The Belgian priest Lemaître had both better and worse luck with his theory of the early universe. It was promptly published, but in an obscure journal and was unknown to the other principals in the field as they made their own discoveries, as remarked above. However, four years later, at a meeting in Pasadena, Einstein proclaimed the importance of Lemaître's work.

The universe that Friedmann had described in his paper was one with a positive curvature, $k = +1$, and a zero cosmological constant, $\Lambda = 0$. Gravity is destined to overwhelm it. It will start from a singularity, when the universe was infinitely dense and hot. Nothing can be said about the contents and conditions at that time with any confidence except for the behavior of the universal scale R. At that singular moment it begins at zero and increases to a maximum value, but the gravity of the whole universe slows the expansion from the very beginning. After the maximum has been reached, the scale of the universe contracts; it shrinks and reheats again after billions of years. This fate is sometimes referred to as the "big crunch".

3.7.2   *Einstein–de Sitter universe*

Willem de Sitter, a Dutch astronomer, and Albert Einstein published a joint paper in 1932 describing a universe that begins with a big bang and expands forever, decelerated at first by gravity, and finally settling into an eternal coasting expansion. It has a flat spatial geometry and a zero cosmological constant and is the simplest of the model universes, as can be seen from the Friedmann–Lemaître equation (page 74), because the last two terms on the right are absent, by assumption. This model is in contrast to the previous one, for which the deceleration eventually becomes zero at a maximum expansion, after which the universal contraction accelerates.

3.7.3   *Accelerating universe*

In the light of recent discoveries, the most interesting model universe is the one that was studied independently by Friedmann and Lemaître, who discovered universal expansion with eventual acceleration as a solution to

Einstein's equations. Evidence for their solution has been growing during the past several years from data gathered with ever more precision on distance and velocity (redshift) of far-off supernovae, on minute details in the cosmic background radiation, and in numerical simulations of the formation of galaxies and galaxy clusters.

All these data point to a flat or nearly flat universe and one whose expansion is *accelerating*. The acceleration, if proven true, will most certainly be the outstanding discovery in cosmology of the past several decades. For it suggests the existence of an unknown form of evenly distributed energy called dark energy. Although its nature is not known — not yet, and maybe not for a long time, if ever — its mere existence is all that one needs to know to gauge its effect on the universal expansion. That effect can be represented by a single parameter, the *cosmological constant*, the term that Einstein introduced into his general relativity, which he later removed, and which is now reappearing as one of the great cosmic mysteries.

3.7.4   *Recycling universe*

The Friedmann–Lemaître equation on page 74 that describes the expansion of the universe is undefined at the beginning of time because the densities of radiation and matter are infinite then. Relativity can give no information whatsoever either about the time before $t = 0$, or the time after the universe has contracted to infinite density again. It is possible that if the universe is actually closed, it will recycle time and again, expanding and then contracting. But we can be sure that, in the details, the past will not be repeated in the future. There is no reincarnation. This follows from the *second law of thermodynamics*, concerning the inevitable increase of entropy (disorder). For the same reason, time does not flow backward after the universe has expanded to its maximum and begins to collapse. That there can be no perpetual motion may be considered fortunate or unfortunate, depending on one's disposition. In this kind of universe, were it not for the second law, we would be destined to live our lives forward and backward, forward and backward, with a 15-billion-year blackout between each direction of living. In any case, since there would be no awareness of the long night between existences, it might become very boring in the long run, not to say alarming at the points in time that we start becoming younger.

3.8 Boxes 5–8

---

5 Einstein Equations

Einstein's equations are $G_{\mu\nu} = -8\pi G T_{\mu\nu} + \Lambda g_{\mu\nu}$, where $G_{\mu\nu}$ is the Einstein tensor, composed of the Riemann tensor, $R_{\mu\nu}$, which characterizes the curvature of spacetime, and the Ricci scalar curvature R. Newton's constant is denoted by G. The presence of matter and radiation appears in the stress–energy tensor, $T_{\mu\nu}$. Einstein's equations tell spacetime how to curve to the presence of matter and radiation, and then tell the latter how to arrange themselves and move under the influence of gravity. It is in this equation that we see that spacetime is not merely an arena in which things happen, but is itself shaped by what happens.

---

6 World Lines and Cosmic Time

Time always progresses, and in one direction only. Therefore, even if a particle sits still, nevertheless, in four-dimensional spacetime it traces a track called its world line. A mathematician, Herman Weyl, hypothesized that world lines of particles in a universe such as ours do not become entangled. (If any two world lines were to cross, the single-valuedness of functions of time would be lost.) In a uniform and expanding universe, an observer could see the world lines diverging from a point at some distant finite or infinite time in the past, but never again would they meet. It is remarkable that Weyl introduced his hypothesis before Hubble had discovered the expansion of the universe. The idealization is altogether reasonable. Of course, sometimes galaxies collide. But, as to the overall history of the universe, we are not disturbed by these events.

If on each world line a common *time t* is marked, the points so singled out form a *spatial* surface. The surface might simply be a plane, in which case the geometry would be Euclidean, like the geometry of lines inscribed on a sheet of paper on a desk. In fact, recent discoveries in cosmology have confirmed that this is indeed the curvature of our universe—flat. It need not have been so. The surfaces at any slice of cosmic time could have been spherical or hyperbolic. The three possibilities emerge as the only curvatures that are possible for the metric of a uniform homogenous universe. Robertson and Walker discovered this metric independently.

7   Metric for a Uniform Isotropic Universe

Express Weyl's hypothesis in terms of coordinates and metric. A world line is labeled by three space coordinates x^m ($m = 1, 2, 3$) and a time coordinate x^0. Consider a 3-surface defined by an orthogonal slice through the world lines at a common time x^0, which we use to label such slices. To satisfy the Weyl hypothesis, the metric tensor g^{mn} must have the following properties. Orthogonality is expressed by $g_{0n} = 0$. Each of the world lines, $x^m = $ constant, is a geodesic. Therefore,

$$\frac{d^2 x^m}{ds^2} + \Gamma_{kl}^m \frac{dx^k}{ds} \frac{dx^l}{ds} = 0 \,,$$

where the line element is $ds^2 = g_{kl} dx^k dx^l$. For $x^m = $ constant (each $m = 1, 2, 3$) we obtain $\Gamma_{00}^n = 0$, and $\partial g_{00}/\partial x^n = 0$. Thus g_{00} depends only on x^0; we can therefore replace it by a suitable function of itself that makes $g_{00} = 1$. The line element then becomes $ds^2 = c^2 dt^2 + g_{mn} dx^m dx^n$, where $t \equiv x^0$ is cosmic time.

1. Example: Surface of negative curvature

$$x_i^2 - (ct)^2 = -R^2 \,.$$

Substitute

$$x_1 = R \sinh \chi \, \cos \theta \,, \qquad x_2 = R \sinh \chi \, \sin \theta \, \cos \phi \,,$$
$$x_3 = R \sinh \chi \, \sin \theta \, \sin \phi \,, \quad t \equiv x_4 = R \cosh \chi \,.$$

This gives

$$dx_i^2 - (cdt)^2 = R^2 [d\chi^2 + \sinh^2 \chi (d\theta^2 + \sin^2 \theta d\phi^2)] \,.$$

Now substitute $r = \sinh \chi$ to obtain

$$ds^2 = c^2 dt^2 - R(t)^2 \left[\frac{dr^2}{1 + kr^2} + r^2 (d\theta^2 + \sin^2 \theta \, d\phi^2) \right] \,.$$

This is known as the Robertson–Walker metric. Here, $R(t)$ is the previously discussed scale factor. For a homogeneous isotropic universe, Einstein's 10 independent field quantities $g_{\mu\nu}(x_\sigma)$ have been reduced to a single function of cosmic time, the scale factor $R(t)$, and a curvature parameter k. The constant k can take three values: $k = 1$ for spherical subspace, $k = 0$ for a planar, and $k = -1$ for a hyperbolic.

8   The Friedmann–Lemaître Equations

For the Robertson–Walker line element derived above corresponding to a *homogeneous and isotropic* universe, only two of Einstein's
field equations are independent. They can be taken as

$$\dot{R}^2 + kc^2 = (1/3)(\Lambda + 8\pi G\rho)R^2$$

and

$$\ddot{R} = (1/3)[\Lambda - 4\pi G(\rho + 3p/c^2)]R\,.$$

Here $\rho = \epsilon/c^2$ is the mass density, ϵ the energy density, p the
pressure, Λ Einstein's cosmological constant, and k the curvature
parameter.

Take the derivative of the first of the above pair of equations,
multiply the second by $\dot{R}$, and eliminate the Λ term from the re-
sulting pair to obtain the conservation law implicit in the Einstein
equations (divergenceless stress–energy tensor),

$$\dot{\rho} = -3(p/c^2 + \rho)(\dot{R}/R)\,.$$

This equation can also be written in two different ways:

$$d/dt(\rho c^2 R^3) = -p\, dR^3/dt\,,$$

which is the energy–work equation for expansion or contraction.
Another way in which the conservation equation can be written is

$$d\rho/dR = -3(p/c^2 + \rho)/R\,.$$

The independent equations governing expansion may be taken as
the first of the Friedmann–Lemaître equations together with the
local conservation equation in any of its forms. This was our implicit choice on page 74.

We can derive rigorously the behavior of radiation and matter
densities that we arrived at by logic in the text from either of
the conservation equations. The equation of state for radiation is
$p = (1/3)\rho_r c^2$. Therefore

$$d\rho/\rho = -4dR/R\,.$$

This yields the conservation equation $\rho_r \sim 1/R^4$. For matter,
$p \ll \rho_m/c^2$ and we obtain instead $\rho_m \sim 1/R^3$. Thus, as we learned,
radiation dominates early in the history, and matter next.

4 Elementary Particles — Fundamental Forces

It remains therefore that the first matter must be atoms....

— Sir Isaac Newton, *Quaestiones*

4.1 The Atom

As recently as 40 years ago there was a branch of physics called *elementary particle physics*. It was an outgrowth of the field of nuclear physics that was pioneered by scientists who had earlier in their careers performed experiments and conceived theories having to do with atomic and nuclear physics: these discoveries had led to the revolutionary quantum theory. For a time they believed that they were exposing the fundamental "atoms", the indivisible particles of which all other things are made. However, the proton, the neutron, and then their antiparticles were only the beginning of a particle catalogue of bewildering size and complexity of interrelations. Evidently none of these was more fundamental than the other. Nevertheless, scientists in the laboratory were tracing, in reverse, the order of things as they appeared in the early universe. So let us begin at the beginning, which is in order of appearance, closer to the end.

In hindsight it is easy to overlook the struggles that preceded major advances in understanding the natural world. The notion that there must be an end to the reduction of matter to smaller fragments is ancient. Perhaps the first to expound this idea was the Ionian Leucippus (480–420 B.C.). His disciple, Democritus (470–380 B.C.), named these ultimately small particles "*atomos*", meaning "indivisible", and we have inherited this word as "atom". It seemed to Democritus that what gave the atoms of each element their distinctly different properties was their size and shape. The actual substances of the natural world were composed of mixtures of the atoms of the different elements, and one substance could be changed into another by altering the mixture. These were certainly prescient notions. The Roman poet Lucretius (95–55 B.C.) was so convinced (with good reason) by these ideas that he wrote a long didactic poem to expound the atomist view.[1] The poem has survived intact, but the early Greek atomists had litle influence on later developments.

Much as we admire the inquiring minds and the pure reasoning that led to the conception of fundamental particles, no understanding of the underlying nature of matter was achieved for two-and-a-half millennia, not really until the last few years. The fascination with pure reason as an instrument for understanding the world had serious limits. It was only when scientists (called philosophers) began to test ideas against observation and experiment that real progress was made. As to fundamental structure, for this we had to await the invention of the cyclotron by Ernest Orlando Lawrence, and the building of the "atom smashers", the Synchrotron and Bevatron at Berkeley, and the Cosmotron at Brookhaven, to reveal how far from fundamental the neutron and proton were.

But even before the need for such power as provided by modern accelerators was recognized, many clues along the trail had to be deciphered. To give some idea of the fog that had to be penetrated, let us turn first to Marie Curie and her discovery of radioactivity.[2]

Marie Sklodowska (Curie) (1867–1934), at that time a young student in Paris freshly arrived from Poland, was looking for a thesis topic. Hints of the divisibility of matter were coming from several quarters. The mysterious X-radiation that was produced when cathode rays struck the glass vacuum tube, like that still used in most television sets and computer monitors, allowed Wilhelm Röntgen to see the human skeleton through the flesh. The discovery of these "radiations" was followed a few months later by Henri Becquerel's "uranium rays", which darkened photographic plates in his closet. Fascinated by these discoveries, Marie boldly chose to search for additional radiations, leading to a lifelong passion for understanding what they signified for the atom, the supposedly indivisible final constituent of matter. Her research eventually won her *two* Nobel Prizes — one in Physics, one in Chemistry. This is a distinction shared by only two other persons in science — Linus Pauling, who won the Chemistry and Peace Prizes, and John Bardeen, who won the Prize for his invention of the transistor (now universally used in electronic apparatus in place of vacuum tubes) and for the theory of super-conductivity (shared with Cooper and Schrieffer).

In the spring of 1894, Marie Sklodowska sought help from a Polish acquaintance to find laboratory space in which to set up her first experiments. He referred her to his colleague Pierre Curie, who had done pioneering

---

[1] Lucretius, *De Rerum Natura* (*On the Nature of Things*).

[2] Mme. P. Curie, *Comptes Rendus*, Vol. 126 (1897), p. 1101.

Fig. 4.1. Madame (Sklodowska) Curie in 1903, the year her doctoral thesis on radioactivity was published in Paris. *Permission: Association Curie et Joliot-Curie.*

research on magnetism, and was at that time laboratory chief at the Municipal School of Industrial Physics and Chemistry in Paris. This meeting forever changed their personal lives as well as the course of science.

About 15 years before his meeting with Marie Sklodowska, Pierre Curie and his brother had invented a new kind of electrometer, which could measure extremely low electrical currents. Instead of using photographic plates to detect radioactivity, an imprecise measure of activity, Marie used Curie's electrometer to measure the faint currents that can pass through air that has been ionized by the passage of uranium rays. Normally, an atom is charge-neutral — it has as many negatively charged electrons surrounding the *nucleus* of the atom as it has positively charged protons in the nucleus. However, very little energy is needed to detach an electron. When this is done, the atom has a positive charge and is said to be ionized. If the atoms are in a gaseous state, the presence of light mobile electrons will conduct an electric current. It was the degree of ionization as measured by the current that Marie Curie used to detect radioactivity. After first confirming

Becquerel's uranium rays, she began a lifelong search for other types of radiation.

She and Pierre Curie, now her husband and colleague, were able to separate other radioactive elements from ore called pitchblende, containing radium and polonium, named after her native land.[3] Three types of radiation were later identified — alpha, beta, and gamma rays.[4] In any case, Marie Curie postulated that the radiations were a property of the atom, and though still not known in what way, the notion of the indivisibility of the atom was beginning to crack. Her own words — written several years after the discoveries — reveal more clearly than I can the mysteries that shrouded the subatomic world:

"Radium is a body which gives out energy continuously and spontaneously. This liberation of energy is manifested in the different effects of its radiation and emanation, and especially in the development of heat. Now, according to the most fundamental principles of modern science, the universe contains a certain definite provision of energy, which can appear under various forms, but cannot be increased [conservation of energy].

". . .If we assume that radium contains a supply of energy which it gives out little by little, we are led to believe that this body does not remain unchanged, as it appears to, but that it undergoes an extremely slow change [the radioactive *half-life*].... . Furthermore, radioactivity is a property of the atom of radium; if, then, it is due to a transformation, this transformation must take place in the atom itself. Consequently, from this point of view, the atom of radium would be in a process of evolution, and we should be forced to abandon the theory of the invariability of atoms, which is at the foundation of modern chemistry."[5]

In the year following her discovery of radioactivity (1897), the next crack came with J.J. Thompson's realization that the cathode rays emanating from the electrically heated wire in a sealed vacuum tube (Figure 4.2, similar in performance to a TV tube) and moving toward an oppositely charged plate, were actual particles that "so far from being wholly aetherial, . . . are in fact wholly material, and that they mark the paths of particles of matter charged with negative electricity".[6]

---

[3] P. Curie, Mme. P. Curie, and G. Bémont, *Comptes rendus de l'Acadamie des Sciences*, Paris, Vol. 127 (1898), p. 1215.

[4] Alpha rays turned out to be the nuclei of helium-4 and beta rays were discovered to be a new type of particle, the electron.

[5] Marie Sklodowska Curie, *Century Magazine* (January 1904), p. 461.

[6] J.J. Thompson, *Philosophical Magazine*, Vol. 44 (1897), p. 293.

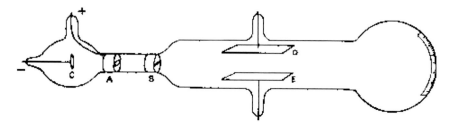

Fig. 4.2. Illustration of a cathode ray tube (similar to a TV tube) from Thompson's paper on measurements concerning electrons and their emission from atoms. "The rays [electrons] from the cathode C pass through a slit in the anode A, which is a metal plug...connected with the earth; after passing through a second slit in another earth-connected metal plug B,...; they then fall on the end of the tube and produce a narrow well-defined phosphorescent patch. [The phosphorescent patch is the site of production of the X-rays discovered by Röntgen.] ...the rays were deflected when the two aluminum plates were connected with the terminals of a battery.... The deflection was proportional to the difference of potential between the plates...." (J.J. Thompson; see footnote 6.)

Thompson continued in his paper to describe how he deduced from the results of his experiments the ratio of the charge to the mass of the type of particle that was set loose in his apparatus, and now known as the *electron*: "What are these particles? Are they atoms, or molecules, or matter in a still finer state of subdivision? To throw some light on this point, I have made a series of measurements of the ratio of the mass of these particles to the charge carried by them." From the experiments he performed, Thompson learned that cathode rays consist of *particles* which are a "subdivision of matter...very much further than [atoms or molecules]...." In fact, the electron which he had discovered is the first really elementary (indivisible) particle in nature. It can be divided no further and, as we shall see, it fits into a pattern with other elementary particles whose discoveries were still many decades away.

Thompson's remarkable paper was correct in all the conclusions relating to the small mass and large charge and other aspects of these constituents of atoms that have come to be known as electrons. However, the known attraction of unlike electrically charged bodies, and the repulsion of bodies of like charge, together with the as-yet-unknown quantum theory, posed a predicament for understanding the structure of atoms. With the knowledge then available, Thompson did the best that he could; he supposed that the positive charge was spread throughout the atom — in what form he did not know (the proton was not yet known) — and that the electrons were

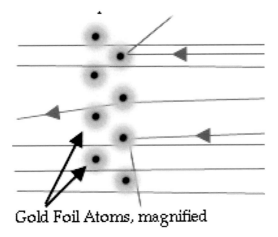

Gold Foil Atoms, magnified

Fig. 4.3. In Rutherford's laboratory at Cambridge, alpha particles were beamed at a thin gold foil, containing of course gold atoms. At the time it was thought that the atom was a mushy mixture of electrons and something having positive charge, and it was therefore thought that the alpha particles would simply pass through the foil with very little deflection. Instead, some alphas were deflected to very large angle while others were not. The conclusion is clearly that an atom has a diffuse halo of electrons surrounding a minuscule but heavy center or nucleus. The figure illustrates the almost straight line trajectories of alphas that miss the nuclei, and the large deflection of those that do strike it. *Permission: "The Particle Adventure", by the Particle Data Group at the Lawrence Berkeley National Laboratory.*

interspersed throughout the entire atom so that positive and negative charges canceled each other out in what is referred to as the "plum pudding" model. The correct resolution had to await many discoveries and theories: the experiments in Rutherford's laboratory; the separate discoveries of protons and neutrons; the genius of Niels Bohr's atomic model; the development of the quantum theory by Schrödinger and Heisenberg; and the revelation of a new force of nature — the strong nuclear force — by Yukawa.

The New Zealander Ernest Rutherford (1871–1937) and the Canadian chemist Frederick Soddy, working together in Canada, came to the revolutionary conclusion that the atoms in the radioactive experiments were themselves spontaneously splitting into the atoms of other elements by emitting one or the other of the two types of radiation, alpha or beta rays (helium-4 and electrons, respectively). The spontaneous decay of one element into another continued until a stable element was at last attained.

After Rutherford had become a professor at Cambridge, an experiment (Figure 4.3) was performed under his direction in 1909 that proved that the atom was nothing like a "plum pudding". Rather, there was a high concentration of mass and of positive charge called the nucleus in a small region at the center of the atom; the nucleus was surrounded at a great distance by shells of electrons of such a number as to exactly balance the charge in the nucleus. The space that an atom occupies is mostly a void, with a massive center and halo of electrons, somewhat like our solar system.

Years were to pass before the actual constituents of the nucleus were discovered; 10 were to pass before Rutherford in his laboratory at Cambridge discovered the proton, and another 12 before James Chadwick, also at Cambridge, discovered the neutron in 1931.

How did it all fit together? It was clear that a positively charged nucleus could hold the negatively charged electrons from flying off into space. However, two deep mysteries were presented by the discoveries of Thompson and Rutherford: Why did the electrons, attracted by the positive charge on the nucleus, not fall in? What kept the protons confined in the small nucleus at the center, despite their mutual repulsion caused by their positive charge? Clearly there must be another force to keep the protons bunched together that is stronger, but of a shorter range, than the electric force.

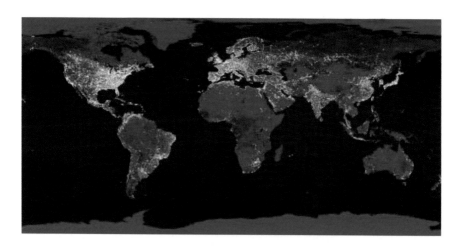

Fig. 4.4. An actual composite photo taken from satellites of the world showing night illumination produced by street and yard lights. The illumination of vapor and dust in the atmosphere obscures the night sky, rendering many stars invisible to city dwellers. *Credit: U.S. Air Force Defense Meteorological Satellite Program (DMSP) and processing by the NOAA National Geophysical Data Center.*

Fig. 4.5. Hideki Yukawa, professor of theoretical physics at Kyoto University, conceived the idea that the *strong* nuclear force, one of the four fundamental forces of nature, is caused by a messenger particle that he called a meson, later discovered and called a pion. The other three are the *weak* nuclear force, the gravitational force, and the electromagnetic force. © *The Nobel Foundation, with permission.*

But the nature and origin of such a force had not yet been discovered. As to why the electrons did not fall down, the explanation of this puzzle was the inspiration for the quantum theory.

Hideki Yukawa (1907–1981), a theoretician in Kyoto, puzzled over these mysteries; by combining relativity and quantum theory he succeeded in describing the attraction between nucleons[7] by a messenger particle (or force carrier) that he called a meson — a new and so-far-undiscovered type of particle.[8] Yukawa (Figure 4.5), who won the Nobel Prize for his insight, was able to infer the approximate mass of the hypothetical particle, the mediator of the attractive nuclear force, from the internucleon spacing in the nucleus.[9] The *pion* was discovered 14 years later. Here, yet, is another example where a mystery was solved by proposing the existence of a particle

---

[7] The generic term "nucleon" is used to refer to either a neutron or a proton.

[8] H. Yukawa, *Proc. Phys.-Math. Soc. Japan*, Vol. 17 (1935), p. 48.

[9] He arrived at a value of 200 MeV, not so far from the mass of the meson, called a pion (140 MeV).

and some of its properties, which only years later was actually discovered. (Other instances are Dirac's antielectron (Section 4.2), Gell-Mann's quarks (this chapter), and Pauli's neutrino (Section 5.4.3).

The constitution of the atom was now known if one probed no further. A central small nucleus contained neutrons and protons, which were bound together by the strong nuclear force. The nucleus was surrounded at a great distance by electrons. The number of electrons exactly balanced the number of protons. As we shall discover later, the number and their arrangement into shells gave to each *element* its distinctive chemical properties. Any element may have some slight variations known as *isotopes*, which differ only in the number of neutrons and therefore the atomic weight of the element.

Now the time was ripe to milk the discoveries of Marie and Pierre Curie, of J.J. Thompson, and of E. Rutherford, and the wealth of spectral data on atoms that had been accumulated over the course of a century for the insights they might yield into the structure of matter.

The Curies had shown that some atoms change spontaneously at various rates into other atoms with the emission of alpha particles or electrons. Thompson's experiments showed that electrons are lightweight charged particles that belong to the atom. Rutherford demonstrated that electrons reside at a great distance from a nucleus having a high concentration of mass and charge at the atom's center. For a century many scientist had revealed that when atoms of a given element are heated, and the light that is given off is passed through a prism, only lines of a few colors are produced by the glowing substance. This contrasts with the continuous spectrum of white light from the Sun that Newton had revealed. Moreover, the pattern of lines for every element is *different* from that of every other element; the pattern is like a fingerprint (Figure 4.6). Counterwise, if white light were shone through a gas of a given element and passed through a prism, the continuous spectrum of colors was interrupted here and there by black lines. The missing colors in the second case corresponded to the only colors that were present in the first case — again a unique fingerprint. Above all, there were regularities in the pattern of electronic energies of many elements that were named after their discoverers, like the Lyman, Balmer, and Paschen series.

The unifying principles underlying these diverse phenomena were utterly mysterious, but the experiments could be repeated time and again with the same results. Line spectra were, and are to this day, of great value in astronomy, for they allow astronomers here on the Earth to learn what elements are present in stars, and in the great clouds of gas in interstellar

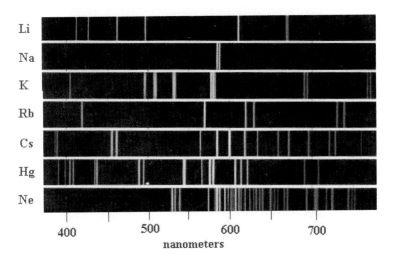

Fig. 4.6. Line spectra of a few elements — lithium (Li), sodium (Na), potassium (K).... Note the two brightest yellow lines of sodium, which produce the characteristic color of street lamps in many cities. The light produced from sodium bulbs is quite suitable for night lighting and does not "pollute" the night sky with light so much as a normal incandescent bulb that produces a continuous broad spectrum of light. City lights produce a glow in the Earth's atmosphere that obscures the view of stars, and generally renders the Milky Way invisible to city dwellers.

space. By comparing the lines of a star's spectrum as shown in Figure 1.14, for example, with those of known elements, as in Figure 4.6, the composition of the star can be determined.

Useful as the line spectra were, how could these observations be fitted together with the discoveries of Thompson and Rutherford, and exploited to penetrate the deeper secrets of the atom? On the one hand, Rutherford had clearly demonstrated that electrons in an atom are at a great distance from the nucleus. But what keeps them there in the face of the attractive electric force between positive nucleus and negative electron? The first thought to come to mind is that the electrons are moving about the nucleus in circles or other closed figures of motion such as ellipses, just as the planets move around the Sun in stable orbits. We know that the tendency of a planet at every instant is to move off in a straight line in the direction in which it is momentarily moving, but that it is deflected by the gravitational attraction of the Sun; that the tendency to move away is exactly countered by the attraction, so that the orbit is stable; it retraces itself time and

Fig. 4.7. The Sun photographed with a filter to view the extreme ultraviolet (171 øA). At the extreme temperature (5000°C at the surface and much hotter within), heat flow produces a turbulent atmosphere. *Credit: Copyright by SOHO Extreme ultraviolet Imaging Telescope (EIT) full-field Fe IX, X 171 A; images from NASA Goddard Space Flight Center.*

again.[10] But this notion cannot be applied directly to the electron, because in the classical theory a charged particle always emits radiation when it is accelerated, as would be so because of the deflection. In such a case, the electron would spiral into the nucleus as a result of the loss of orbital energy to the creation of radiation photons — not at all a stable situation for the electron. Evidently, a gap in the knowledge of physical laws existed — a gap that was to be filled with the discovery of the quantum theory.

At the time that Bohr tackled these obstacles, there was great unease among the world's leading physicists. Bohr approached the problem against this background with a keen awareness that certain inadequacies

---

[10]Planetary orbits appear to be stable on the timescales on which we are able to observe them. However, we know from Einstein's theory of gravity that the orbiting planets and moons must all radiate gravitational waves, which are disturbances of spacetime. Thus all orbiting bodies will gradually lose orbital energy and the orbits will shrink. For planets the timescale is very long, and for the Earth's moon, the rate at which energy is lost in raising tides is greater. However, for closely orbiting neutron stars, like the Hulse–Taylor pair, the effect of gravitational wave radiation has been measured and confirms Einstein's theory to a very high accuracy.

were showing up in the classical theory of electricity and magnetism. Planck had found it necessary to suppose that light was emitted by a perfect source called a black-body in *quantized* amounts of energy, a notion that many of his colleagues doubted but that Einstein had proven was true in an even more real sense by his explanation of the particlelike behavior of light in ejecting electrons from the surface of metals (Section 2.3). Indeed, light, whose wavelike properties were well known, nonetheless behaved as packets of energy (photons) in this circumstance; light has particlelike properties alongside its wavelike character.

de Broglie's insight into the *wavelike* character of *particles* was still in the future.[11] But much the same notion was contained in Bohr's theory of the atom.[12] In fact his quantization of electron orbits in hydrogen can be equivalently stated as requiring that any electron orbit should have a circumference that is equal in length to an integer number of de Broglie wavelengths of the electron.[13]

However, Bohr puzzled over his strange idea for some time after he left Rutherford's laboratory to return to his native Denmark. There, a friend asked him what his theory would say about atomic spectra and the Balmer formula, which was a simple empirical formula that characterized the wavelength of light emitted or absorbed in a transition between two atomic states of hydrogen. Johann Balmer was a teacher in a girl's school in Basel, who published only three papers in his lifetime; two of them, written at the age of 60, were on the spectrum of hydrogen. Bohr was unfamiliar with the work until his friend referred him to it. His response was quick: "As soon as I saw Balmer's formula the whole thing was immediately clear to me."

Bohr's theory of the atom was not perfect and its rationale was *ad hoc*. It accounted for some but not all of the spectral lines that had been observed in hydrogen and other atoms and studied intensely for a century (Figure 4.6). However, the signposts set by Planck, Bohr, Einstein, and de Broglie were well directed. Heisenberg, Schrödinger, and Dirac quickly followed with the quantum-mechanical theory — the theory of mechanics that must be applied to the motion of the very small.

---

[11]See page 36.

[12]Niels Bohr, *Philosophical Magazine*, ser. 6, Vol. 26 (1913), p. 1.

[13]Bohr's quantum hypothesis for the electron orbit is that the orbital circumference must be an integer multiple of its wavelength $\lambda = h/mv$. Bohr's quantization condition reads $C = 2\pi r = n\lambda$, where r is the orbital radius of the electron, n is any integer beginning with 1, h is Planck's constant, m is the electron mass, and v is its velocity.

A new and deeper understanding of atomic structure emerged from this quantum theory of the micro world. Now the different chemical bonding properties of elements can be understood in terms of the arrangement of electrons into different shells as their number increases from one element to another. This is so important in chemistry, and therefore in innumerable ways related to the universe and our own lives, which we pause to explain.

Electrons obey what is called the *Pauli exclusion principle*. Pauli realized that certain particles (now often referred to as fermions) have the very special property that not more than one of them can be in a given quantum state of motion. The consequence of this law for the structure of an atom is that the electrons are arranged in many different quantized orbits, according to the number of electrons and the *quantum numbers* of the atomic orbits.

We have encountered the notion of quantization before, in connection with emission and absorption of light that produces line spectra by which the composition of solar atmospheres can be discerned. Many kinds of particles have an intrinsic property called their *spin* angular momentum (which is always specified in the units of Planck's constant, $h/2\pi$). Those with a half odd-integer spin, like $1/2$, as for the electron, proton, and neutron, are called fermions, and it is they that obey the Pauli principle. Apart from this intrinsic property that belongs to the particle at all times, the quantum states of an electron in an atom are characterized by the orbital angular momentum, and by the way the orbital angular momentum and spin are added together to give their total angular momentum. There is also a quantum number that specifies the orientation of the orbit, and one that characterizes the electron motion in their quantum orbits that, roughly speaking, denotes their distance from the nucleus. So altogether an electron has a number of *quantum numbers* that are characteristic of its motion in the atom.[14]

All of these quantum states with the same angular momentum and differing only in orientation have the same energy, and all these electrons form a *shell*. According to Pauli, only one electron can have any one set of these four quantum numbers, which also specify the electron energy in the orbit. From this it follows that while most of the electrons of the atoms of different elements are in common states of motion, one at least is in a different state, which sets each element apart from all others. There is more to be said about a shell of electrons; it is quite literal as quantum mechanics goes. The Heisenberg uncertainty principle does not permit the precise

---

[14]They are called $n, l, j = l \pm 1/2, m = -j - 1/2, -j + 1/2, \ldots, j + 1/2$, of which for each n, l, j there are $2j + 1$ different states of motion which characterize the orientation of the angular momentum.

spatial location of an electron in its quantum state (see Section 2.4). So the electrons of an atom occupy fuzzy shell-like regions around the nucleus.

Thus had the pursuit of the elementary or fundamental "atom" led to one of the two great revolutions in 20 century physics — quantum mechanics, the strange mechanics of the microworld (relativity being the other). But of the three particles discussed so far — the electron, proton, and neutron — only the electron is elementary.

4.2 Vacuum: Particles, Antiparticles, and Dirac

> Because...you have believed since childhood that a box was empty when you saw nothing in it, you have believed in the possibility of a vacuum.
>
> — Blaise Pascal, *Pensées*

Paul Adrian Maurice Dirac[15] was born in 1902 in Bristol, England and lived his last years in Tallahassee, Florida, where he died at the age of 82. His childhood was austere, and so was the rest of his life. When he was a child his father, born in Switzerland, insisted that only French be spoken at the dinner table; Paul was the only child to do so, in consequence of which he and his father ate their meals in the dining room while the wife and other siblings ate in the kitchen. Their upbringing was so strict that the children were alienated in an unhappy home.

From his first year in school his exceptional ability in Mathematics became clear to his teachers. When he was 12 years old he entered secondary school, attending the school where his father taught, the Merchant Venturers Technical College. He completed his school education in 1918 and then studied Electrical Engineering at the University of Bristol.

Dirac took the Cambridge scholarship examinations in 1921 and won a cholarship to St John's College, but had insufficient funds to support himself. Instead, he accepted an offer to study Mathematics at Bristol without paying fees and he was awarded first class honors in 1923. Following this he was given a grant to undertake research at Cambridge and he began his studies there in the same year.

Ralph Fowler, the leading theoretician in Cambridge, supervised Dirac's research. Though Dirac would have preferred to do research in general relativity, Fowler steered him to statistical mechanics and quantum theory. He completed five papers in two years. Otherwise, it was a difficult period;

---

[15]Most of the information here was gleaned from an article by J.J. O'Connor and E.F. Robertson.

his brother committed suicide and he became estranged from his father. He was a solitary person with few friends, and withdrew even further after this tragedy.

Dirac frequently took walks in the country, and on one of these he was struck by certain implications of Heisenberg's uncertainty principle.[16] This led him to formulate for the first time a mathematically consistent general theory of quantum mechanics in correspondence with Hamiltonian mechanics.[17]

Upon receiving his Ph.D. degree in 1926, Dirac went to work with Niels Bohr in Copenhagen, and then on to Gottingen in February 1927, where he interacted with Robert Oppenheimer, Max Born, James Franck, and the Russian Igor Tamm. There, Ehrenfest invited him to spend a few weeks in Leiden before returning to England. He was elected a Fellow of St John's College, Cambridge at the age of 25. Three years later he was elected a Fellow of the Royal Society. Dirac made the first of many visits to the Soviet Union in 1928. In 1930 he published *The Principles of Quantum Mechanics*, an austere and much-loved treatise noted for its abstract approach, relying hardly on experiment. Perhaps for this reason the book is ageless, and as well read now as then. A link between quantum mechanics and relativity, first made in a short paper, is repeated there.

Dirac was appointed Lucasian Professor of Mathematics at Cambridge at the age of 30, a post he held for 37 years. This was the same post held earlier by Sir Isaac Newton. In the year following this appointment, he received the Nobel Prize for Physics, which he shared with Schrödinger. He lived a long and fruitful life, and is commemorated by a plaque in Westminster Abbey. Stephen Hawking, who was Dirac's successor in the Lucasian Chair, presented the memorial address.

Shortly after Einstein discovered special relativity, Dirac formulated a relativistic theory of the electron by requiring that the quantum theory of the electron should obey the special theory of relativity. It is an amazingly simple theory, yet Dirac's equations suggested something quite unexpected — an *antielectron*. It was soon discovered in cosmic rays, and given the name "positron". This was the first time, but far from the last time, that theoretical physicists have been able to predict the existence of new particles from theory before they were actually discovered.

---

[16]Dirac realized that Heisenberg's uncertainty principle was a statement of the noncommutative property of the quantum-mechanical observables.

[17]Dirac's 1930 book *Quantum Mechanics* remains a treasured classic.

The antiproton was discovered much later because its mass is about 2000 times greater than the mass of an electron, and therefore required a high energy accelerator that had to await the genius of Ernest Lawrence and his invention of the cyclotron (and the construction of the Bevatron in Berkeley). A host of other discoveries followed, and it appears to be a law of nature that for every particle type there is a corresponding antiparticle with precisely the opposite properties, such as positive versus negative electric charge of the positron and the electron.

When a particle and an antiparticle of the same type come within a short distance of each other, they annihilate each other; pure energy appears in their place carried by a photon of light, X-ray, or gamma ray, depending on how small or great the mass of the particle is. All the matter in the universe consists of a great many more particles than antiparticles. If they were equal in number, the universe would have been annihilated particle by antiparticle. Does this particle–antiparticle asymmetry have any explanation in the context of the creation of the universe? Perhaps.

The discovery of antiparticles led to another discovery — vacuum fluctuations, which are related to the Heisenberg uncertainty principle. This principle acts at the quantum level where Heisenberg found that nature does not permit both the position *and* the momentum of a particle to be measured as accurately as an experimenter might wish (see Section 2.4). Likewise, nature does not permit infinitely accurate measurement of the mass of an unstable particle (like the neutron). The accuracy of an experiment designed to measure mass and lifetime, M and τ, is limited. If the mass is measured to within an accuracy ΔM, then the lifetime measurement will have an uncertainty of $\Delta \tau$, where that uncertainty is governed by the law

$$\Delta M c^2 \Delta \tau > \hbar,$$

and vice versa, with $\hbar$ being Planck's constant (divided by 2π). Plank's constant is of such a size that in everyday life we do not notice the limitation. The reader need not expect to understand *why* this is a *law* of nature. It is a consequence of quantum mechanics, which we make no attempt to explain here. So we must accept it, and by doing so we can understand some of the other peculiarities of nature at the submicroscopic or quantum level.

Another law of nature assures that energy is conserved. The implications of these two laws, the uncertainty principle and energy conservation, are quite surprising. For example, vacuum is empty spacetime, meaning that no particles are present in the region of the vacuum. But does this mean the region — say, the inside of a box — is really empty of everything? No! The uncertainty principle and the reality of antiparticles imply that the vacuum fluctuates — that particles and antiparticles in pairs are forever appearing for a short time and then disappearing. The mass of the pair and the duration of their existence before they mutually annihilate are related by the above law of Heisenberg. Such a pair is referred to as a *virtual* pair, meaning not real in the sense that each pair must disappear after a very short time that is dictated by the above law. The greater the particle–antiparticle mass — which, from Einstein's theory, we know is equivalent to energy ($E = mc^2$) — the shorter the time for which each pair can exist.

The consequence of the uncertainty principle is that the vacuum is filled with virtual particle–antiparticle pairs that flicker into existence and out again. Can any sense be made of this seemingly preposterous notion? Does it have observable effects on things that can be measured?

It may seem strange that, out of nothing, particle–antiparticle pairs are briefly appearing and disappearing everywhere about us. Our senses are too coarse to detect the small effects associated with fluctuations in the vacuum. However, accurate experiments using delicate equipment can and has measured the fleeting appearance of particle–antiparticle pairs. Dirac's relativistic theory (with which he predicted the possible existence of antielectrons called positrons) can be used to calculate very accurately the quantum energy states of the simplest of all atoms, namely the hydrogen atom, just because it is the simplest atomic system — one electron bound by the electric force to one proton.[18]

Nevertheless, a small discrepancy was discovered between the calculation and the superbly accurate measurements of Willis Lamb. He won the Nobel Prize for measuring the effect of the transitory electric currents of vacuum fluctuations on certain energy levels in hydrogen atoms. The discovery was essential to the development of particle physics in the latter half of the 20th century. It confirmed the reality of the vacuum fluctuations that theorists expected from their quantum field theory, and paved the way for understanding nature at its most fundamental microscopic level — the quarks and gluons, about which we will have more to say later.

This is the Lamb shift: A hydrogen atom consists of one proton and one electron and therefore is electrically neutral, and no electric field would be observable from outside the atom. However, on the very small scale of the atom itself, the continuous and random appearance and disappearance of the electric charge of virtual positron–electron pairs creates a weak

---

[18]Refer to Figure 1.15 for the meaning of *quantum states*.

and highly variable electric field which slightly disturbs the motion of the electron in its orbit around the proton of the hydrogen atom; certain of the quantum energy levels of the atom are shifted slightly by the disturbance. The very small shift in the energy levels that is caused by the virtual appearance and disappearance of positron–electron pairs was actually calculated to be in extremely close agreement with that measured by Lamb in 1947. This is inescapable evidence that the vacuum, rather than being empty, is like a sea, roiled by currents, crosscurrents, and eddies — the result of the constant creation and annihilation of all sorts of particle–antiparticle pairs. Vacuum fluctuations may also act on the largest scale (the scale of the entire universe), causing the rate of expansion of the universe to *accelerate*.

Fluctuations of the vacuum may have created small black holes in the very early universe. The gravitational field is so strong just outside a black hole that energy can be transferred from the black hole to the particle–antiparticle pair of a vacuum fluctuation sufficient to draw one of them into the real world. This is a quantum effect and it leads to a quantum correction to Einstein's theory of gravitation. The correction is completely unimportant except for very small black holes. It was realized by Bekenstein and Hawking that black holes have properties reminiscent of those of hot bodies, in which case they might radiate, contrary to the classical concept of a black hole. Indeed, Hawking calculated the rate of the quantum-mechanical process of evaporation of a black hole by absorption of antiparticles and liberation of particles in fluctuations of the vacuum. In this way, small black holes, whose temperatures are very high according to Bekenstein, evaporate very rapidly, even explosively, if small enough.

4.3   Antiparticles and Antiuniverses

Some scientists have speculated that vacuum fluctuations played a crucial role in the process of creation of the universe. I think it is entirely possible that there is no deep scientific explanation (in the sense, say, of fundamental symmetries) of why there are more particles than antiparticles in our universe. Perhaps there are two, or many universes, created out of nothing in which particle and antiparticle numbers simply balanced out; that our universe is part of one big vacuum fluctuation; that the reason it has survived so long is simply that it is so improbable that the antiuniverse of ours would have evolved in precisely the same way — to have the same configuration of all its parts — so that the two can mutually annihilate; that the more time that elapses the more unlikely that the right combination of configurations of the parts will ever occur. This is a safe bet unless someone discovers a deep scientific explanation for particle–antiparticle asymmetry in our universe. But this paragraph is unverifiable speculation and we return now to science.

4.4   The Particle Explosion

The dream of Democritus had come true. The chemistry of all substances could now be understood in terms of atoms, of which there were many different types. Each atom consisted of electrons bound in quantum states by the electric force to a small nucleus of opposite charge formed from protons and neutrons held tightly together by the strong nuclear force. Substances owed their differences in part to the mass of the nucleus but especially to the exclusion principle and the number and arrangement of the electrons.

But the question why some nuclei spontaneously undergo radioactive decay by emitting alpha, beta, or gamma rays remained a puzzle that physicists set about investigating by inventing and constructing nuclear accelerators. The need for accelerators to attain a high energy of motion for their projectiles — protons, alpha particles, or electrons — was evident to them from de Broglie's discovery that particles have also a wavelike character with a wavelength related to their mass (m) and velocity (v) by the famous de Broglie relation $\lambda = h/mv$, involving Planck's constant, h.[19] For an experiment to be sensitive to details of size d, a "probe" — an accelerated proton, for example — has to have a wavelength that is smaller than d; that is to say, it has to have a high velocity or, equivalently, high energy. We encountered the same idea in discussing how the part of the spectrum that is useful for vision must have wavelengths that are similar to the size and spacing of sensory cells in the retina (Figure 2.9). Thus, as time went on, and finer detail was needed to probe the small distance structure of particles, ever-higher energy accelerators were called for (Figure 4.8).

4.4.1   *Cosmic rays*

Meanwhile, other physicists were intrigued by the serendipitous discovery of cosmic rays by a Jesuit priest and physicist. In 1910 Theodor Wulf climbed the Eiffel Tower with a device that he had himself designed and built to detect energetic charged particles from radioactive minerals. He found that his device counted a greater rate for the passage of particles at the top of

---

[19]See Section 2.3.

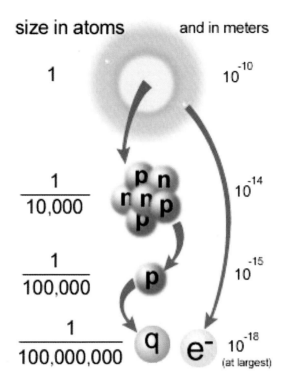

size in atoms   **and in meters**

1   10^{-10}

$\dfrac{1}{10,000}$ 10^{-14}

$\dfrac{1}{100,000}$ 10^{-15}

$\dfrac{1}{100,000,000}$ 10^{-18}
(at largest)

Fig. 4.8. Relative sizes of subatomic particles as compared to an atom, its nucleus, nucleons, quarks, and electrons. *Permission: Particle Adventure, Particle Data Group, Lawrence Berkeley National Laboratory.*

the tower than at the bottom. This was not what he had expected; all the radioactive substances previously studied had been found in mining ore. He proposed that balloon experiments be flown to discern if the count rate continued to increase in the rarefied atmosphere above the Earth, and this was found to be so. Eventually, these experiments uncovered the existence of new particles similar to the proton as well as new families of particles that had been produced by high energy particles of cosmic origin colliding with atomic nuclei in the atmosphere. Among the products was the pi *meson*, the particle that Yukawa had predicted as the carrier of the nuclear force. The antielectron, called the positron, was also discovered in cosmic rays. Its discovery confirmed Dirac's prediction of antiparticles from his relativistic quantum theory of the electron (Section 4.2). And then another surprising discovery — a heavy electron called the muon.

Cosmic ray physics experiments tapped cosmic accelerators of unimaginable power, giving protons an energy 100 million times the energy achievable by the most powerful Earth-bound accelerators. Although many exciting and unexpected discoveries were made, cosmic rays have also clear disadvantages compared to a controllable laboratory environment that can provide a particular probe — say, electron or proton — with a specified energy and on an orderly time schedule.

4.4.2   *Laboratory beams of particles*

Machines for accelerating beams of such particles as electrons, protons, and later nuclei were invented, the most successful of which was the *cyclotron* of Ernest O. Lawrence. As the name suggests, the particles that are to be used as the probes are cycled round and round in the machine and given a boost in velocity in each cycle. They start from a source at the center of the machine (Figure 5.4). Perpendicular to the path that they are directed on is a strong magnetic field created by the poles of a very large and heavy magnet. A charged particle moving in such a field is deflected to the side, so the path of the charged particle — say, a proton — is constantly being deflected; instead of flying out of the machine, it circles within it. Twice during each cycle, the proton passes a gap across which an electric field has

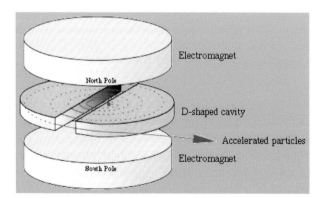

Fig. 4.9. Schematic illustration of the *cyclotron*. The dashed line marks the path of charged particles from where they are injected at the center and travel enlarging orbits because of the boost they receive on passing from one D-shaped chamber to the other, across which an oscillating electric field is applied. The whole machine sits between the poles of two circular magnets that deflect moving charged particles. This curves the path of the beam so that it recycles time and again to be boosted twice each turn.

been applied; this boosts its velocity by acting on its charge, and naturally increases the radius of the orbit.

As was expected from the principles of electrodynamics, the time taken for the particle to travel around the machine in its new expanded orbit was the same as in all the previous times; therefore, a beam of such protons could be boosted in velocity and hence in energy when an oscillating electric field of constant frequency was applied at a fixed location in the circular orbit of the protons. This permitted the acceleration of not one proton, or a bunch of them, but a continuous beam, because no matter where in the cycle of acceleration they were located, they always arrived at the right time at the right place to be boosted to a higher energy. The first such machine, built to verify that the concept actually worked, was no bigger than a toy. Now the largest one is five miles across. It is buried in an underground tunnel near Geneva and is wide enough for a spacious pathway for maintenance of the magnets and 15-mile-long vacuum ring in which particles are accelerated (Figure 4.10).

4.5 So Many Riches

Scientists with their photographic plates on mountaintops to detect cosmic rays, scientists flying their apparatus in balloons for the same purpose, scientists in laboratories with their powerful cyclotrons, all in search of the fundamental particles. What emerged from all these activities was confusion. The proton and the neutron along with the electron had been thought of as elementary constituents of the atom. Instead, countless other particles of a similar nature to the proton and the neutron were discovered as well as unrelated particles — more than 200 new particles through the 1950s and early 1960s. It became clear that none of these was more fundamental than the other.

So many riches became an embarrassment, as it became clear that this zoology of particle species comprised *composite* particles — particles made of something still more fundamental or elementary, just as the electron and the nucleus are more fundamental than the atom. The appellation *elementary particle physics* was changed to *high energy physics* or, simply, *particle physics*. Only recently has the original term won recognition again.

Nevertheless, amidst all the confusion, certain very basic laws were distilled from the sea of data. For example, certain relations among different particles were observed. One group of particles had the special property that as many as were present before a reaction reappeared among the final products. In contrast, other particles appeared for the first time only

Fig. 4.10. A cross section of the LEP accelerator ring at CERN, Geneva. The ring is 5 miles in diameter and the tide of the Moon causes this to stretch and shrink daily by as much as 1 millimeter, as can be sensed by the accuracy of the detectors used in experiments (Figure 5.4).

after a reaction. Particles of the type whose number was conserved were called *baryons* and the others were called *mesons*. Neutrons and protons are baryons, and the pion, among many other particles, is a meson.

A typical reaction that revealed the new conservation law is denoted in shorthand by the expression

$$n + p \rightarrow 2p + \pi^- + \pi^0 \, .$$

This is the first time an expression of this type has been written, so let us pause to discuss its meaning and usefulness.[20] Reading from the left, it signifies that a neutron and a proton collide, and the arrow indicates that from this collision the particles on the right emerge. It is understood that the total energy, comprising the energy of motion of the particles and the energy equivalent of their masses, is the same before and after the collision, in accord with conservation of energy; likewise for their momentum. Ever since Maxwell discovered the mathematical laws governing electromagnetism, physicists have been convinced that the number of units of electric charge, e, reappears in the final products as was initially present; that electric charge is conserved. This can be verified by counting charges on each side of the above reactions.

Still other reactions were discovered in which antibaryons appeared among the reaction products, such as

$$n + p \to 2n + p + \bar{p} + \pi^+ + \pi^0 .$$

The antiproton is denoted by the bar, hence $\bar{p}$. Many reactions, like the above two, demonstrated a new conservation law: just as many baryons, no more, no less, always emerge after a reaction as were present before. An antibaryon cancels a baryon in the count. The law of *conservation of baryon number* was recognized. Whatever these particles carried within themselves, their *net* number was conserved. There were as many baryons and whatever it takes to make baryons in the early universe as there are now.

Of course, such a strong statement, that the universe contains as many baryons *now* as near the *beginning* of time, requires much stronger evidence than discussed so far. Indeed, scientists have obtained such evidence, and in the usual way that evidence of very rare events is sought. Rather than watching a single proton to see if it decays — quite impossible if its life is longer than ours — one tracks what happens to a huge number of them to see if one of that huge number decays. This is what is done in the Super-Kamiokande experiment in Japan (Figure 4.11). An enormous detector containing seven hundred thousand billion billion (7×10^{32}) protons was built to monitor their fate. Not one proton has been observed to decay. The present experimental evidence is that the proton lives longer

---

[20]We know already that the neutron carried no charge and that the proton and the electron carry a positive and a negative unit of charge, respectively. Otherwise, the charge on a particle is indicated by a superscript; thus π^+ denotes a positive pion.

Fig. 4.11. Super-Kamiokande detector in Japan, half full of pure water, with photon detector tubes in view above the water line to register the fleeting flashes of blue light marking the trails of rare cosmic events. When full it will contain 50 000 tons of water, which contains 7.5×10^{32} protons. If the proton lifetime were 10^{32} years, then about 7 proton decays should be detectable in a year. Taking account of the efficiency of the counters in detecting decays, and the length of observation time (longer than a year), it was found that the lifetime exceeds 10^{33} years. *Permission: Institute for Cosmic Ray Research, The University of Tokyo.*

than 1.6×10^{33} years, which is vastly longer than the present age of the universe (1.5×10^{10} years).[21]

Another important clue to the substructure of baryons and mesons emerged from the observation that certain baryon types were always made in association with a meson called the kaon; for others, no such rule held. For example, the kaon does not appear among the products of any of the reactions written above, but it appears in reactions in which a lambda (Λ)

---

[21] Shiozawa *et al.*, *Physical Review Letters*, Vol. 81 (1998), p. 3319.

or a sigma (Σ) baryon is made:

$$n + p \rightarrow \Lambda^0 + p + K^0 ,$$
$$n + p \rightarrow n + \Sigma^+ + K^0 .$$

Those baryons that were made in association with kaons were called "strange". It was indeed a mystery and worthy of the adjective "strange" until some years later, when Murray Gell-Mann, Y. Ne'emann, and George Zweig noticed that baryons and mesons could be arranged into something like Mendelyaev's periodic table of the elements. But, in the above two reactions, how were the lambda and the sigma identified as the particles that carried baryon number, rather than the kaon? The answer became quite clear when the kaon was observed to *decay*:

$$K^0 \rightarrow \pi^+ + \pi^- .$$

This symbolism indicates that after some unspecified time, the kaon disappears and its energy, charge, and momentum are conserved by the oppositely charged pions that are produced. Because the number of pions is not conserved, the kaon does not carry the conserved baryon number; rather, it is the Λ^0 and the Σ^+ that are the strange baryons.

4.6 The Quarks and Leptons

Gell-Mann's classification of baryons and mesons had associated with it what were initially thought of as fictitious particles called *quarks*. They were very peculiar because, according to the rules by which the baryons and mesons were made from quarks, the quarks had fractional charges. Unlike other charged particles such as electrons and protons which had one unit of electric charge e, and nuclei which had an integer number of units, like $1e, 2e, \ldots$, some quarks had $2/3\,e$ and others $-1/3\,e$. Fractional charges were just not supposed to appear. Baryons, like the proton and the neutron, were made up of three quarks, while mesons, such as the pion, were made up of two, a quark and an antiquark. These structures are in accord with another property attributed to quarks: they each carry $1/3$ units of baryon number. Thus three quarks make a baryon, and a quark and an antiquark a meson, which is not a baryon because it has no net baryon charge. But a quark can never be liberated from the nucleons and mesons that contain them. They are said to be *confined*. Still, as we will see, they are real constituents of what must be regarded as composites, the baryons and mesons.

Fig. 4.12. Murray Gell-Mann (in 1983), winner of the Nobel Prize for his discovery of the classification of baryons and mesons that led to the greatest revolution in the theory of elementary particle physics. *With permission of Murray Gell-Mann.*

Two types of quarks are called "up" and "down" (for want of another name). A third is called "strange", because it always occurs in the baryons that were earlier called "strange". The kaon is always made in any reaction that makes a strange baryon; therefore, a conservation law must be at work — that for strangeness. Strangeness is conserved in these (strong) interactions, such as the last two reaction equations written above. The strange baryon Λ must have one strange quark and the K one antistrange quark, because in the reaction above there were no strange quarks initially. By observing many reactions such as those written above, the quark content of the neutron, proton, lambda, and kaon were thus deduced to be

$$n = (udd), \quad p = (uud), \quad \Lambda = (uds), \quad K^0 = (d\bar{s}),$$

and the charges carried by the quarks to be

$$u = 2/3\,e, \quad d = -1/3\,e, \quad s = -1/3\,e.$$

Likewise, the quark structure of the some 200 other baryons and mesons were deduced.

If, in any of the reactions written above, one counts the net number of baryons, of electric charge, and of strangeness, before and after the reaction, one will find them unchanged — all save for the *decay* of the kaon itself, in which strangeness has disappeared (as in the decay interaction shown at the end of Section 4.5). However, an important observation was made concerning the speed of the reactions and decays: strangeness *is* conserved only on the very short nuclear scale of time (the time taken for all the *reactions* above is 10^{-23} seconds). However, after a longer time (but still incredibly short, like 10^{-10} seconds) the kaon decays. This fact revealed an important difference in the forces involved. Evidently there is a *strong* nuclear force at work in the fast reactions and a *weak* nuclear force in the slow. And *strangeness* is conserved only in strong interactions, but not in weak. So, strangeness number disappears almost as soon as it is made.

The transformations at the quark level of the decay of the neutral kaon, K^0, into a positive and a negative pion are shown in Figure 4.13 in what is called a Feynman diagram. There are rules based on quantum field theory by which the decay probability can actually be computed from the details of such diagrams. That is to say, the theoretician knows what mathematical symbols are to be placed in an equation corresponding to each line and

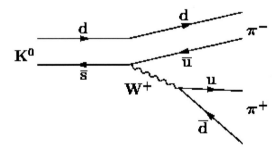

Fig. 4.13. The Feynman diagram for the decay of a neutral kaon. A K^0 is composed of a down quark d and a strange antiquark $\bar{s}$, shown at the left. Time progresses to the right. The antistrange $\bar{s}$ quark is transformed through the carrier of the weak nuclear force W^+ into an anti-up-quark $\bar{u}$ to form, with the d quark of the K^0, a pion π^-. The W^+ boson itself shortly decays into a u and $\bar{d}$ quark to form a positive pion π^+. Together, the final products carry the original (zero) conserved electric charge *and* the original (zero) baryon number, but *strangeness* has disappeared through the intervention of the carrier of the weak force, the W^+.

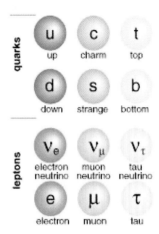

Fig. 4.14. The fundamental particles of all matter. Six quarks flavors, each with three colors, six leptons of which three carry negative charge (including the electron) and three chargeless neutrinos. Each has an antiparticle.

vertex in such a diagram, after which it remains for one to carry out the computation indicated by the equation.

This counting of charges and so on may seem like some kind of a numbers game, but there are more subtle organizing properties that are all too intertwined for the notion of quarks, at least as mathematical entities, to be ignored. One final property of quarks needs to be mentioned. Whether a mathematical fiction or real, quarks cannot ever appear in isolation. Now it really sounds like a con game. "They are there, but you cannot see them." Gell-Mann was of course quite aware of how extremely peculiar this all seemed and for several years he was quite equivocal about whether he thought of quarks as being "real" particles, or convenient mathematical fictions by which hadrons could be classified and the rules governing their transformations in reactions summarized.[22] It is not at all apparent from what is written here how profound the classification really is, for it involves a field of mathematics called *group theory*.

With the success of the quark model in making sense of the large array of what are now known as composite particles — the baryons and mesons — searches were made for the quarks. Some searched for the quarks themselves as free objects; however, never has an isolated quark been found, and it is

---

[22]*Strange Beauty: Murray Gell-Mann and the Revolution in 20th-Century Physics*, by George Johnson.

generally believed now that a quark cannot exist as a free particle, separated from others. Each quark type — the up quark, for example — has three variations, which are called "color". It is a postulate that an object with (this kind of) color cannot be observed; that observable particles must be made from such a combination of quarks that is colorless, in the same sense that white light is a mixture of all colors, but is not itself a color. Thus a meson is made up of a quark and an antiquark of the same color. Between them, color and baryon number cancel out, as should be so because mesons are not baryons. Baryons are also colorless because they contain one quark of each of the three colors.

The first experiments that revealed the quark structure of protons were carried out at the Stanford Linear Accelerator for very high energy electrons. The setup was very much like Rutherford's scattering experiment which revealed the presence of a very small nucleus inside an atom (recall Figure 4.3). The high energy electrons, behaving as a very small probe, in accordance with de Broglie's wavelike view of particles, revealed finer structural details inside the proton; Richard Feynman and James Bjorken were able to interpret the internal structure as being due to very small scattering centers which they called *partons*. Partons are in fact the quarks, the parts of baryons. Aside from the three light quarks, there are heavy quarks called "charm", "truth", and "beauty" (sometimes "top" and "bottom"), denoted by c, t, and b, which were discovered much later (the top as recently as 1995).

It is truly amazing how, after 2500 years, the existence of one class of the *fundamental* particles — the quarks — which can never be individually seen, was nonetheless divined with the help of the conservation laws, deduced as illustrated above, and by the powerful organizing principles of an abstract field of mathematics called group theory. Murray Gell-Mann, Yuval Ne'emann, and George Zeig were the magicians. Gell-Mann, because he contributed to every aspect of the development of the quark structure of baryons and mesons, won the Nobel Prize.

Electrons share a similar conserved attribute to baryons and quarks called *lepton* number, but they share it with a distinctly different particle of zero electric charge and very small or zero mass called the *neutrino*. Each has its antiparticle, and the lepton number can neither be created nor destroyed. Evidently, the universe contains a fixed number of these *conserved* particles — the quarks and leptons — that existed near the beginning and would endure to the end (or almost).[23] Behold the linkage between the microcosm and the cosmos.

---

[23] See footnote 21.

4.7   The Force Carriers

Force is such a powerful and mysterious concept. Bring a bar magnet near some iron filings and they are arranged in a peculiar pattern at the poles. Lose hold of a stone and it falls to the Earth. An electron is attracted to the positively charged proton. The nuclear force holds the atomic nucleus together and can *transform* a proton in collision with another proton into a neutron and a charged pion. A neutral kaon spontaneously disappears and in its place two pions of opposite charge appear. We have names for the forces: gravity, electromagnetic force, the strong nuclear force, and the weak nuclear force.

To have a name is something, but not much. These forces act across space. Gravity acts over the extent of the entire cosmos. So it has a long range, though in comparison with the others it is very weak. The nuclear force is very strong, but acts only over a very short distance, no further than the distance between nucleons in a nucleus (10^{-13} centimeters). What is really meant by "acts" and *why* does one act over a great distance and another over a small, subnuclear distance? Can any rhyme or reason be attached to these facts? At one time it was believed that forces could act instantaneously over a distance through the aether (later realized to be an artifact). Besides being illogical, we now know through the principles of quantum mechanics and quantum field theory that the fundamental forces in nature are carried by particles that are referred to as *force carriers* and that no transmission can take place at a greater speed than light.

The force carriers belong to the class of unconserved particles called bosons. It is an unusual thought as far as everyday life is concerned — that force is carried by a particle. Where are they; where do they come from? We are so accustomed to the experience of gravity that we may think of it as being just there. The force carriers go about their work in *virtual* states of existence. This means that they go about their work under circumstances in which the energy that is required for their *real* existence is not actually there. However, according to the Heisenberg uncertainty principle, energy and time cannot be simultaneously and precisely defined: the product of the uncertainties in energy, ΔE, and in time, Δt, are related by the Planck constant by

$$\Delta E \Delta t \sim h\,.$$

This is the quantum uncertainty that limits the range of fundamental forces and of unstable particle lifetimes. Essentially, the above equation says that energy of no more than ΔE can be borrowed from the vacuum for a length

Fig. 4.15. An instrument called a bubble chamber, which registers the passage of charged particles by creating small bubbles along the particle track, shows in this case an antiproton ($\bar{p}$) hitting a proton in a nucleus (not visible, but located at the point where a shower of particles originates at the center left). Positive and negative pions are created in equal numbers, thus conserving the total charge of the colliding proton–antiproton pair in this reaction. At the bottom left, a π^+ decays into a muon μ^+ and a neutrino (which, because it is neutral, does not leave a track). The creation of a neutrino is apparent because of the conservation of momentum: the muon moved to the right of the direction of the disappeared pion, so the neutrino has moved to the left. The tracks of the charged particles are curved because of the presence of a magnetic field that is perpendicular to the plane of the picture. Positive-charged particles are deflected in one direction; negative in the other.

of time not exceeding $\Delta t \sim h/\Delta E$. In that short time, a light signal could travel a distance $r = c\,\Delta t \sim hc/\Delta E$; consequently the force range cannot exceed this distance. Because the mass of the force carrier is related to energy by $E = mc^2$, we learn that the greater the mass m of the force carrier, the shorter the range of the force. Yukawa estimated in this way the mass of the nuclear force carrier, later to be uncovered as the pion.

The expectation that the ranges of forces are inversely proportional to the mass of the force-carrying particle has been fully confirmed in experiments. How can this be done? We have just shown that the range of the force carrier is limited by the uncertainty principle because of the absence of energy to make the particle real under most circumstances. But the experimenter, with powerful accelerators such as LEP at Geneva, *can* supply the energy to make a force carrier real, and thus to prove its existence and measure its mass and electric charge. In the five-mile-diameter LEP accelerator, a beam of high energy electrons is made to collide with a similar beam of positrons (antielectrons) coming from the opposite direction. According to the conservation laws, the energy can be converted into the mass of a particle having zero electric, baryon, and lepton charges (because these *conserved* charges are absent to begin with). In this way one of the force carriers of the electroweak interaction, the Z^0 , was first detected. The Z^0 itself is not a stable particle and lives only for a very short time. Using the uncertainty principle in a similar way to that above, the lifetime of the Z^0 can be determined by the small but finite range of energy for which production of the Z^0 is successful. We will encounter this *reaction width* of the unstable Z^0 again on page 133 and learn how it provided the first laboratory determination of the number of neutrino flavors. At the same time we will see that the apparatus was so sensitive that it also measured the tide raised on the Earth's crust by the Moon.

Just as the photon (γ) is the force carrier of the electromagnetic force, the Z^0 (together with the $W^\pm$ bosons) are the carriers of the weak nuclear force. However, electrons interact, not only through the massless photons, but also through the very massive bosons of the weak nuclear force, while neutrinos interact only through the weak force. Actually Sheldon Glashow, Abdus Salam, and Steven Weinberg unified the two forces under the name of the electroweak force independently in 1967. They shared the Nobel Prize in 1979. The force carriers, very massive $W^\pm$ and Z^0 bosons, were not discovered for four more years.

Quarks, now understood as real particles but not able to exist in isolation, are the basic constituents of baryons and mesons. What holds them together in the baryons and mesons? Gluon is the name given to the bosons that carry the strong force that acts between quarks. There are eight types of them. This number is related to the deeper group-theoretic mathematical structure into which the *fundamental* quark particles fit. Gluons are denoted by g. They are analogous to the photon (γ), which is the electric force carrier between charged particles. However, there is one difference between gluons and photons that has a profound effect. Photons do not carry electric charge, but gluons carry color charge. Therefore, gluons can interact among one another as well as with quarks, whereas photons can only interact with electrically charged particles. This interaction of the force carrying bosons among themselves is what makes physical problems involving the strong nuclear force so difficult in comparison with electrodynamics.

Because the photon has zero mass, the range of the electric force is very long. In contrast, the range of the weak force, the one for example that destroys the strangeness by which the kaon decays, such as

$$K^0 \rightarrow 2\gamma \,,$$

is accomplished through a heavy force carrier, the Z^0. The range of both forces is in accord with the Heisenberg principle and therefore is very short.

The force between quarks is the most curious of all the fundamental forces. Unlike the others, it becomes stronger the greater the distance between a pair of quarks. This is in accord with the observed fact that the color of strong interactions cannot be seen by itself, but only in a colorless state such as achieved by baryons through a three-quark structure of all three colors, or by bosons, through a quark and an antiquark of the same color. Why color cannot be observed is unknown. However, the mathematical theory of quark and gluon interactions must have this observable fact embedded in its structure, and indeed does so. The theory is called *quantum chromodynamics*. Together with the electroweak theory, it constitutes what is called the *standard theory of particle physics*.

5 The Primeval Fireball

The evolution of the world can be compared to a display of fireworks that has just ended: some few red wisps, ashes and smoke. Standing on a well-chilled cinder, we see the slow fading of the suns, and we try to recall the vanished brilliance of the origin of worlds. . . .

— Georges Lemaître, *The Primeval Atom*

5.1 Cosmic Evolution

Before there were stars or galaxies, the early universe blazed with fire; it was filled with a blinding light together with electrons, protons, neutrons, neutrinos, mesons, and all their antiparticles. Now, there are great voids between galaxies, but then, everything in the cosmos — everything within our horizon — occupied a very small space. The density of light, of particles and antiparticles, and the temperature of the universe were much higher than found in any object in the heavens today. The temperature inside the Sun is more than *10 million* degrees on the Kelvin scale[1]; at the center of a neutron star at birth, it is *500 billion* Kelvin. Yet, with confidence, we know what the contents of the universe would have been at even much higher temperatures.

How is it possible that we can name the contents and trace the evolution of the universe from a time beginning at a small fraction of a second following its birth in the Big Bang? The laws of nature enable this marvelous feat. Here is how. Even though the universe does not have a fixed temperature — it drops as it expands — the rate of expansion is so slow compared to the rate at which light quanta were emitted and absorbed

---

[1] On the Centigrade scale, water freezes at $0°$. From the human perspective, this is a useful standard. However, zero ought to have a more profound significance than the freezing temperature of water. On the Kelvin scale, zero degrees, which is denoted by 0 Kelvin and is called the absolute zero, denotes the complete absence of heat; there can be no lower temperature. The interval of one degree has the same meaning in the two scales. Otherwise, 0 Kelvin equals $-273.160°$C.

by nuclei and atoms, that a type of equilibrium existed, nonetheless; it is called adiabatic equilibrium. This is important because the powerful laws of thermodynamics and statistical mechanics developed in the late, 1800s to the early 1900s and the particle discoveries during the 1950s–'60s in large accelerators at Berkeley, Brookhaven, and later at CERN, Geneva, can be used to trace the sequence of events that took place even at the particle and atomic level as the universe expanded and cooled. All that has been learned about atomic, nuclear, and particle physics in laboratories throughout the world is needed to trace the detailed history of our early universe.

Philosophers and scientists as early as Kant and Newton realized that the universe had not been created as it is now, but evolved from some earlier form. However, this view was not accepted generally until Hubble observed the recession of the galaxies and therefore the expansion of the universe in 1929. Newton understood the motion of the Moon and the planets in terms of his gravitational force, but because the stars appeared to be fixed, he conjectured that the universe must be infinite; otherwise gravity would cause it to collapse on its center.

Einstein, when he turned his attention to the universe, also believed that as a whole it was static, though not necessarily infinite, and he briefly introduced what is called the *cosmological constant* into his theory of gravity so as to *prevent* the universe that his theory described from collapsing. Very recent discoveries suggest that he was correct to choose the freedom allowed by his equations of general relativity to add the cosmological constant. But not to keep the universe from collapsing; rather, to account for the apparent cosmic *acceleration* of its expansion.

"How did it all begin?" is a question children ask themselves when their minds begin to grapple with questions of origins beyond their family circle. Beginnings have always posed perplexing philosophical problems. So have endings. The Belgian priest and cosmologist Georges Lemaître was evidently the first to conceive, already in his Ph.D. thesis of 1927, of a fiery beginning of the universe, and the possibility of an eventual accelerating expansion. However, this second part of his work was largely ignored in his own day and forgotten by our time. Indeed, acceleration of the expansion was a great surprise when it was discovered several years ago by Saul Perlmutter, and if it proves to be true, will be one of the major cosmological discoveries of all time, ranking with Hubble's discovery of universal expansion, and with Penzias and Wilson's discovery of the cosmic background radiation that has pervaded the universe since it was a mere 300 000 years old (see Section 5.4.6).

The accumulated evidence for this beginning is now overwhelming. Perhaps it also settles, once and for all, the question of what came before. The moment when the universe emerged in the Big Bang separates what came before and what came after so thoroughly that no logic, no science, and no mathematics can penetrate it. Many scientists, including Einstein, believe that it was a brief moment of such intensity that the laws of physics as we presently know them did not hold; a moment when the laws of the quantum world of the small and the world dominated by gravity met; a world of "quantum foam", as John Archibald Wheeler calls it. As to endings, we are still left to contemplate, if we wish — and as the present measurements of the crucial quantities suggest — a seemingly boundless future. . . .

5.2   Heat, Temperature, and Equilibrium

Although heat in Newton's day was thought to be some mysterious unseen fluid that could move about, we now understand that heat corresponds to the energy of random motion of the molecules, atoms, and electrons of a gas, or of vibrational motion around their normal positions in the case of a solid substance.

Atoms and molecules contain electrically charged particles — protons in the nucleus and electrons surrounding it. In the course of their random motion, atoms and molecules collide. The jiggling of the electrical charges converts some of the energy of motion of charged particles into radiation.[2] It is this radiation that we feel as warmth from the Sun and see as light. In both cases, the radiation has the same nature; it consists of very small packets of energy, called *photons*. Temperature is a measure of the *average* energy associated with the random motion. The higher the temperature, the more energetic the photons and the motion of electrons, atoms, and molecules.

Although the energy of photons is very small, the number arriving from the Sun and falling on a patch, say, of one square centimeter each second is enough to power processes that we experience and witness and use. It is enough to activate electrical signals in the retina of our eyes that are conducted to the brain, where the signals are processed to provide images of the outside world. It is enough to power chemical reactions, like photosynthesis in plants, the only organisms that manufacture their own food, and thereby support the entire chain of life.

When a glass is filled with cold water and left to sit on a table, the heat energy will flow from the hotter room to the colder water until they share

---

[2]See Section 2.5.

the energy in an average way. When the energy is shared among all the modes of jiggling motion and photons, all parts have the same temperature and are in *thermal equilibrium*.

A state of thermal equilibrium is not a static state; far from it. The heat energy is constantly being shifted about among the various modes in which heat energy is carried. But all modes that can be brought into play by a given amount of heat energy in the body share the energy in an average way according to specific laws of thermodynamics as derived by Bose and Einstein in the case of photons, and by Fermi and Dirac in the case of electrons and nucleons.

The rate at which the universe cooled was slow in comparison with the rate of heat exchange among its very dense neighboring parts, so that all the parts of the universe remained in a state of quasiequilibrium, and the laws of thermodynamics can be used to trace its early thermal history. Let us look now at the fascinating variety of processes that took place to shape the universe as it is now, starting with the earliest time we can imagine.

5.3 Planck Era ($t < 10^{-43}$ Seconds)

A time known as the *Planck time* marks the earliest possible time *after* which we can apply the laws of physics as discovered by experiment. Time earlier than that is called the *Planck era*. How is it possible that we can define such a time? Recall (Section 2.8) that Einstein was the first to recognize that light — or, in general, all radiation — has not only wavelike properties but also particlelike properties, and that de Broglie realized soon afterward that particles shared this duality. He found that a particle of momentum p can also behave like a wave with wavelength $\lambda = h/p$, where h is Planck's constant. From Hubble's great discoveries — the expansion, isotropy, and homogeneity of the universe — we have understood that at one time, long ago, the entire part of the universe that lay within any horizon occupied a very small region.

How small could that region be and still be large enough for the laws of physics as we know them now to be valid? We can find the dimension of a region that is certainly *too* small. The normal meaning of space and time in quantum mechanics and relativity cannot hold within a region whose horizon lies within its own de Broglie wavelength. This very early time, called the Planck epoch, refers to that instant during which, as John Archibald Wheeler (Figure 5.1), the renowned nuclear-physicist-turned-relativist, put it, spacetime was so entangled as to merit the name "quantum foam".

Fig. 5.1. John Archibald Wheeler, whose early work with Niels Bohr still provides a very useful way of understanding certain nuclear properties, turned later to relativity and cosmology. He coined the name "black hole" when he became convinced that a strange mathematical solution of Einstein's relativity could really exist in nature, and "quantum foam" to describe the roiling condition of spacetime at the Planck scale.

How can we find out the duration of such a unique situation? The Friedmann–Lemaître equation (page 74) governs how the universe expanded from an intensely hot and dense beginning, and how the expansion was resisted by the self-gravity of the mass–energy in the universe. The solution of that equation will provide the universe's density as a function of time with which we can then solve for that early instant when the horizon of the universe lay within its own de Broglie wavelength. We prefer this physical definition of the Planck epoch (Box 12) to the usual definition, which simply combines fundamental constants in such a way as to yield a quantity with the dimension of time. It is expressible in terms of three of the fundamental constants of nature — Planck's constant, Newton's gravitational constant, and the speed of light — as $t_P \sim \sqrt{hG/c^5}$. Inserting the values of the constants, the Planck time is $t_P \sim 10^{-43}$ seconds. This is such an incredibly small increment of time that we have no apparatus capable of measuring it. Indeed, no apparatus can possibly do so. Atomic clocks have an accuracy of better than 10^{-14} seconds per second.[3] The stability of the period of rotation of some neutron stars rivals atomic clocks. Yet they come nowhere near being able to measure such a short interval of time as the Planck time. Nevertheless, general relativity is a valid theory back to such a time. At earlier times and complementary distances, space and time as we know them have no meaning. Quantum gravity, a union of quantum mechanics and relativity, will reign supreme. But such a union has not been achieved.

Therefore, the Planck time marks that instant *after* which the presently known laws of physics *begin* to apply, and therefore marks the earliest instant we can consider with our present knowledge of physical laws.[4] We therefore want to know what the temperature and density of the universe would have been at the Planck time (Box 12). We summarize these quantities in a table:

$t_P \sim 10^{-43}$ seconds

$l_P \sim 10^{-33}$ centimeters

$T_P \sim 3.6 \times 10^{32}$ degrees Kelvin

$m_P \sim 8 \times 10^{-7}$ grams

$\rho_P \sim 8 \times 10^{92}$ grams per cubic centimeter

During the Planck era the temperature and density were far larger than can be found anywhere in the present universe. In the densest of stars — neutron stars — the density is only $\rho \sim 10^{15}$ grams per cubic centimeter. This is already very high, higher than any earthly machine could compress material, higher than the density of the Earth, which is about 5 grams per cubic centimeter. But at the earliest time the density of matter was 10^{87} times greater than that of the densest stars.

5.4 Radiation-Dominated Era ($t = 10^{-11}$ Seconds to 10^6 Years)

The laws of particle physics including the strong and weak interactions gelled at a time of $1/100$ billionth of a second. It has been discovered in

---

[3] An atomic clock will stay within one second of true time for six million years.

[4] Some cosmologists are seeking to formulate a theory in which gravity, quantum mechanics, and elementary particle physics are unified. It is hoped that such unification will provide a means of probing to even earlier times and perhaps even find a meaning to our vague notions of beginnings and of time itself.

experiments, and confirmed time and again, that baryon number, lepton number, electric charge, and total momentum and energy in all forms are conserved in any reaction. These laws are of great importance to our story of the early universe. They are called *conservation laws*.

Although we cannot recreate the conditions that existed in the early universe, with the conservation laws and that very general branch of physics known as statistical mechanics, we can name the particle contents of the universe, even the proportions of particle types, and still later, nuclei and atoms, as they formed in the cooling universe.

The earliest time at which we begin our detailed story of the evolution of the cosmos is about 1/100 billionth of a second after the Big Bang.[5]

$$10^{15} \text{ K} \quad \text{at about} \quad t = 10^{-11} \text{ seconds}.$$

At that moment we can say literally that there was fire of almost unimaginable intensity. The fire consisted of radiation in its most energetic forms, the terrifically heavy gauge bosons, and the massless high energy gamma rays of light, together with matter in the form of quarks and antiquarks, electrons and antielectrons, neutrinos and antineutrinos (of all three flavors and in some sense all the particles and antiparticles we know now, both those that actually exist in the everyday world and those we know only because of their fleeting existence in high energy collisions in earthly accelerators and in cosmic rays). All of this can be referred to as radiation, even the particles and antiparticles, because all were fluctuating into and out of existence; a photon disappearing into a particle–antiparticle pair, then reappearing with the disappearance of a pair. Almost all the energy in the early universe was in the form of this radiation. The fraction of particles in excess of the number of antiparticles was very small.

All these particles, antiparticles and photons, interact with each other at the speed of light. But they are so closely packed in the dense early universe that the time between collisions among themselves is very short compared to the time it takes for the density to change appreciably due to the universal expansion. Though particles move much slower than photons, the frequent changes in the direction of motion of photons caused by their collisions kept the material particles and radiation together.[6] The

---

[5]The temperature was equivalent to the top quark mass ($\approx 2 \times 10^{15}$ K), the degeneracy factor was $\alpha = 427/8$, and the time would have been $t = 8 \times 10^{-12}$ seconds (Boxes 10 and 11).

[6]The Sun is much less dense than the early universe, yet the time it takes a photon to reach the surface from the core is measured in millions of years, so hindered are they by collisions with electrons which deflect their movement.

whole ensemble, consisting of electromagnetic radiation, electrons, protons, hyperons, neutrinos, all their antiparticles, and ionized atoms, is called a plasma. The entire universe, at early times, was in thermal equilibrium, all the while expanding. Under these circumstances, the expansion is said to be adiabatic.

As the universe expanded, it also cooled. It had to cool drastically before atomic nuclei could be produced. Nuclei are bound systems of neutrons and protons in approximately equal numbers. They are bound together by the strong attractive nuclear force. Being bound means that the mass energy of the nucleus is less than the mass energy of the individual neutrons and protons that it contains. When nucleons approach closely enough, the short range of the nuclear force can hold them together; the difference in energy is radiated as photons that move away at the speed of light. With that energy gone, conservation of energy guarantees that the bound nucleus cannot *spontaneously* decay (fall apart).

However, when the temperature is high (which, recall, means that the total energy — of photons and of randon particle motion — is large), any nucleus that is formed is as quickly destroyed by the bombardment of high energy photons and nucleons. As the temperature dropped because of the expansion, lower energy photons of X-rays and ultraviolet replaced the early high energy photons called gamma rays by the processes of scattering off protons and electrons. The universe had to cool to a temperature equivalent to less than the energy that binds together the simplest nucleus before that nucleus (and more complex ones built from it) could survive the bombardment. When this condition prevailed, the first stable nuclei in the universe were formed.

As the universe cooled further, the continued formation of nuclei would depend on an intricate chain of circumstances, the most remarkable among then involving the *number* of neutrino types and the time between neutrino reactions with other matter which outruns the age of the universe at a very early stage. When this happens, the supply of neutrons needed to form nuclei will soon run out; nuclear fusion then ceases; the numbers of the light nuclei in the universe become frozen at their early values. No more of them would be made, and being stable, and with the temperature decreasing, nothing would destroy them. These are the *primordial* nuclei. The entire supply of light nuclei — deuterium, helium, and lithium — was formed in the first few minutes except for negligible amounts of helium made in stars much later. All the elements heavier than lithium were made much later in the fires of massive stars — about ten sun masses or more — starting at a time of about 800 million years.

Radiation and matter ceased to behave as a single cosmic fluid moving together with the expansion when the agitation caused by heat subsided sufficiently that electrons became bound by the electric force to protons, deuteron, and helium nuclei to form *charge-neutral atoms*. At that point no free electrically charged particles remained, and the radiation photons, which interact most strongly with free charges, were no longer hindered by collisions. Thermal equilibrium between matter and radiation therefore ceased to exist. From that time onward to the future — even to our present time and beyond — the photons would stream freely through the expanding universe. Very few of them would ever interact with matter over these 15 billion years. Those photons, untouched since that early time but much reduced in energy by the expansion of the universe, form the cosmic background radiation, a relic from the distant past. This stage of decoupling of matter and radiation commenced when the universe was 300 000 years old, and was complete by a million years.

Even after decoupling, the mass density of radiation remained greater than that of matter a while longer so that radiation was still the dominant term in the Friedmann–Lemaître equation and controlled the expansion until it faded in importance at about a million years, as we calculate on page 151. During that time, the epoch of radiation dominance, the universe passed through a number of stages, including the formation of the light elements. In the next few subsections let us trace the intricate sequence of events that took place in the first million years to see how strongly those events in the early stages of the Big Bang are reflected in the universe as it is today, 15 billion years later.

5.4.1   *Superradiant era (t = 10^{-11} to 10^{-5} seconds)*

In about 10^{-11} seconds (1/100 000 000 000 seconds) the universe had cooled to about 3×10^{15} degrees.[7] It contained very little that would be recognizable in the world today. The universe was uniformly filled with radiation, consisting of gamma rays which are very-high-energy photons (much higher than the photon energies of light), together with certain particle–antiparticle pairs. Pairs of many types were present — neutrinos which have no mass or very little, electrons and muons (which belong to a family of particles called leptons), and, very notably, quarks and gluons. The nucleons, from which all atomic nuclei today are made, could not have existed

---

[7]Unless it is said otherwise, temperatures will be stated in degrees Kelvin. At such high temperatures as prevailed in the early universe, it hardly matters whether Centigrade or Kelvin units are meant, because they differ only by 273°.

at these temperatures. They are first formed from the quarks and gluons when the temperature falls below about 1000 billion degrees (10^{12} degrees). Nucleons cannot exist at a higher temperature than this.

The universe was a seething, bubbling inferno in which individual gamma rays would disappear for a moment, leaving in their place a particle–antiparticle pair of any of the types mentioned. Any of the particles or antiparticles, upon meeting an antiparticle or a particle of the same type, would annihilate, producing a gamma ray. Between these events, gamma rays moved at the speed of light, and all material particles moved at close to that speed, although they could not move far before being interrupted in their flight. Though the universe was expanding all the while, the rate of expansion was much less than the rate at which the above creation and annihilation processes were taking place. The universe was in thermal equilibrium among all its constituents, both radiation and matter. And there was much more energy density in radiation than in particles. This marked the beginning of the era dominated by radiation.

5.4.2 *Hadronic era (t = 10^{-5} to 10^{-3} seconds)*

Amazing events occurred in the next few moments. At about 10^{-5} seconds (1/100 000 seconds) or a temperature of 3×10^{12} degrees, almost all the quarks were annihilated by their interaction with antiquarks. Strangely, the numbers of quarks and antiquarks were not precisely the same. Otherwise, there would have been total annihilation of matter. But a *relatively* minuscule number of quarks and gluons remained. They combined in an astonishing number of combinations and excitations to form different kinds of exotic nucleons, hyperons, and mesons. This marked the end of the quark era and the beginning of the *hadronic* era (see Figure 5.2).

The era when the exotic nucleons were formed to when they decayed lasted for a few 1/1000ths of a second. Near the end of that era, the exotics decayed into ordinary neutrons and protons, which constitute almost all the mass of the visible universe today.[8]

We know that that brief exoticism existed because one by one hyperons and mesons have been very fleetingly recreated in large particle accelerators like the former Bevatron in Berkeley and the cosmotron at Brookhaven. It takes a small book to catalogue all their names and sketch their properties.

---

[8]The emphasis is on *visible*, the ordinary matter of stars — or, in other words, baryon matter; we do not include in this term the mass of dark nonbaryonic matter, whose nature is unknown. Such matter is important in other connections and will be discussed later.

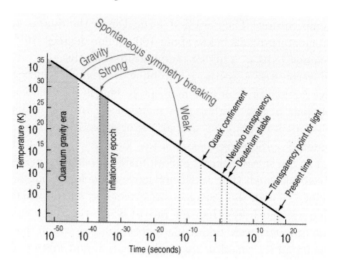

Fig. 5.2. Temperature (vertical axis) of the universe as a function of time from the beginning (horizontal axis). The part of this figure that depends on Einstein's theory is the temperature T as a function of time. The other labels involve other areas of physics that come into play as the temperature falls.

Yet, if they can be created by our feeble means in the laboratory, how much easier and more plentiful their production in the intense fire of the first fraction of a second. For the physicist, it is merely a matter of moments to calculate the temperature above which any particular particle of known mass would have occurred naturally.

Toward the end of the hadronic era, when the most important interactions where the strong nuclear and Coulomb interactions, the leptons and antileptons annihilated, only a small excess of electrons together with neutrinos and antineutrinos remained. These, together with a small excess of neutrons and protons over their antiparticles that were left over from the superradiant era, are like a few grains of sand left high on the rocks after the storm has passed and swept all else away. These few remnants formed the beginning of all we now see — the stars and planets and life.

One of the great unsolved mysteries is why there was a slight surplus of particles as compared with antiparticles. Otherwise all particles would have annihilated and the universe would have been entirely empty save for the ghostly glow of the cosmic background radiation. *That* is all that would have remained from the Big Bang. This marvelous universe and all the enormous diversity that surrounds us never would have been. But there *was* a surplus, and we *are* here to trace our history to the beginning of time.

5.4.3 *Leptonic era (t = 10^{-3} to 1 second)*

> The discovery of...radioactivity adds...to the great number of invisible radiations now known, and once more we are forced to recognize how limited is our direct perception of the world which surrounds us.... .
>
> — Marie Sklodowska (Curie), *Century Magazine*

The neutron mass is greater than that of the proton and the electron together, so in today's universe the *free* neutron is unstable and *decays*: the weak interaction called beta decay transforms a neutron (n) into a proton (p) and an electron (e) in about 10 minutes.[9] These words can be written as

$$n \to p + e + \bar{\nu} \, .$$

The last particle, the antineutrino ($\bar{\nu}$), was undetected in early experiments and remained so for many years. Electrons and neutrinos are among the particles known collectively as leptons. The electron interacts through its electric charge with other charged particles. Otherwise, electrons and neutrinos interact very weakly with matter. They do play a very important role in light element production in the early universe, as we will see.

In the early days of nuclear physics, the French physicist Henri Becquerel, a Polish student in Paris, Marie Sklodowska (Curie), who later was to win two Nobel Prizes, and the New Zealander Sir Earnest Rutherford at the Cavendish Laboratory, discovered that the *atoms* of certain heavy elements, which once were thought to be indivisible, apparently released some sort of emanation because they caused photographic plates to darken. Marie Curie named the unknown phenomenon "radioactivity".

Soon, three types of radioactivity of uranium and other heavy atoms were discovered, and because the nature of the three emissions was not known at once, they were called alpha rays, beta rays, and gamma rays (the first three letters of the Greek alphabet). It was only later that beta rays were discovered to be electrons. For this historical reason the reaction written above is referred to as the "beta decay" of the neutron. Alpha rays were later discovered to be the nuclei of helium-4. Gamma rays were the only true rays of the three; as we learned previously, they are the highest energy photons of the electromagnetic spectrum.

---

[9]Neutrons that are bound in nuclei or in neutron stars are stabilized by their binding energy. It is the half-life that we quote as 10 minutes. This means that in any sample of neutrons, about 1/2 of them will spontaneously decay in 10 minutes.

During the early days of nuclear physics research, the only particles that were observed when a neutron decayed were the proton and the electron. Neither the neutrino nor the antineutrino was known, nor were they discovered for many years; this led to a serious puzzle. The measured total energy of motion plus the mass energy of the neutron did not match the total *observed* energy carried by the proton and the electron into which the neutron had been transformed. Something seemed to be seriously amiss with the law of conservation of energy. Yet, this law was so well established that in 1930 Wolfgang Pauli postulated the an additional particle was created in beta decay reactions. He believed this because otherwise energy was not balanced; it must have very little or no mass but it carries the excess energy (and momentum and spin) but no electrical charge because that was already balanced.

A few years later, the postulated particle was named the neutrino by Enrico Fermi; its existence, even though not yet detected, was taken quite seriously. Fermi developed the theory of radioactive beta decay, assuming the reality of neutrinos as an integral part of the theory. Finally, 26 years after Pauli's neutrino hypothesis, Fred Reines and Clyde Cowan detected the neutrino in a carefully planned experiment at the Savannah River nuclear power plant.[10] Such a site was chosen because it was by then well known that if the neutrino did exist, it must react very weakly with matter. In fact we now know that a neutrino will usually penetrate the entire Earth without interacting. As a consequence an intense source was needed to increase the chance of detection by laboratory instruments; Fermi and others knew that if the neutrino did exist, it would be produced in great numbers in the fission reactions that take place in nuclear power reactors. An experimenter can compensate for a small probability of interaction of *a single* neutrino by using a beam of *a large number* of them passing though the target for *a long time*. For the discovery Reines won the Nobel Prize in 1995.

Years after the discovery of the neutrino, and with the development of a better understanding of that elementary particle, both theoretical and experimental, it became clearer why the experiment of Reines and Cowan was so difficult. The average interval between one interaction of a neutrino with a nucleon, and the next, became *longer than the age of the universe when the universe was only about one second old*. For that very reason, these elusive particles were crucial in limiting the amount of hydrogen in

---

[10]C.L. Cowan, Jr., F. Reines, F.B. Harrison, H.W. Kruse, and A.D. McGuire, "Detection of the free neutrino: a confirmation", *Science* 124, 103 (1956). Frederick Reines and Clyde L. Cowan, Jr., "The neutrino", *Nature* 178, 446 (1956).

the early universe that was converted to helium and several other light elements. The universe would be a very different place now had the neutrino interaction been stronger.

What does it mean to say that the time between reactions of a neutrino is longer than the age of the universe? It does not claim that no neutrino will ever interact with matter in the universe. After all, Reines won his Nobel Prize for detecting a few of them. Moreover, many scientists and engineers have devoted years of their lives to the design, building, and making observations using enormous detectors like Super Kamiokande in Japan (Figure 4.11), and "telescopes", of the most imaginative design, like the AMANDA[11] mission at the South Pole, where holes are drilled about a mile into the Antarctic ice to hold long chains of detectors. The detectors do not detect the neutrinos as such. Rather, the whole diameter of the Earth is the matter with which some few neutrinos may interact to produce heavy electrons (muons) that are charged and *can* be detected. Other large neutrino detectors are the SNO[12] detector, a huge container of heavy water (D_2O) in a deep mine at Sudbury, Canada; and the ANTARES project in the depths of the Mediterranean Sea.

What it means to say that the time between reactions of a neutrino is longer than the age of the universe is that of the billions of neutrinos flying through the universe, only occasionally will one of them interact with another particle. When the physicist calculates a rate, it is an average. Some neutrinos will take much longer, some much shorter, but most around the average. The detectors located in such various places as cited have as their goal the detection of neutrinos from rare astrophysical events. They provide yet another view of the universe, not the optical view of telescopes, nor the radio view of large radio antennae, nor an X-ray view or the gravitational wave view of LIGO, but a *neutrino view*. All are windows on the cosmic events in the universe that record different but complementary information.

Let us unravel the events that put neutrinos in the crucial role of determining — out of all the protons and neutrons present at early times — how many would end up in forming the universe's lifetime supply of the very light elements like deuterium and helium. Neutrons are as essential as protons for making nuclei; all nuclei have at least as many neutrons as protons, and most a few more. However, any neutrons that were not stabilized by being bound in these nuclei decay after about 10 minutes, as discussed above. There is therefore a small window in time when nuclei can be formed

---

[11]Antarctic Muon and Neutrino Detector Array.

[12]Sudbury Neutrino Observatory.

in the early universe. The mass of helium in the universe as compared to the mass in all the rest of matter in the form of baryonic matter has been very accurately measured as being close to 25%. This fraction is extremely sensitive to the evolution of particle types in the early universe, to how long neutrinos continued to interact strongly with the rest of matter in the universe before flying off at nearly the speed of light, and, finally, to the *number* of neutrino types (called flavors).[13]

Flavors in particle physics were explained in Chapter 4. Until quite recently the number of types of neutrinos was not known from experiments performed in earthly laboratories. It is therefore remarkable that the processes governing the formation of helium in the first few minutes were so well known that the number of neutrino types could with certainty be declared as being equal to three. The number three was subsequently confirmed through difficult experiments performed at a giant particle accelerator at CERN, Geneva in 1989.[14]

The object of the experiment, indeed the purpose for which LEP[15] at Geneva was built, was to discover, if they exist, the force carriers of the electroweak interaction that had been predicted by the theory developed in the period from 1961 to 1967, primarily by the work of Sheldon Glashow, Steven Weinberg, and Abdus Salam.[16] The charge-neutral force carrier, the Z^0, was the first of the three predicted particles to be discovered. The experiment provides an example of the Heisenberg uncertainty principle and a particular feature in the data revealed the startling sensitivity of the measurements: it uncovered a gravitational effect of the Moon on the CERN accelerator (Figure 4.10). First, the uncertainty principle and how it revealed the existence of no more than three neutrino families, which, as we have seen, interact with matter more weakly than any other known particle: they usually pass through the Earth as if it were not in their paths. The Z^0 particle has an enormous mass as nuclear particles go. It is 90 times heavier than a proton and corresponds to an equivalent temperature of 10^{14} degrees and it would have occurred naturally in the universe at times less than 1.2×10^{-8} seconds (Box 11).

---

[13] Apparently Peebles was the first to realize that primeval nucleosynthesis might depend on the number of neutrino flavors: P.J.E. Peebles, *Physical Review Letters*, Vol. 16 (1966), p. 410.

[14] "First evidence that the number of light neutrinos = 3", G.S. Abrams *et al.*, *Physical Review Letters*, Vol. 63 (1989), p. 2173.

[15] Large Electron Positron collider.

[16] Glashow, Weinberg, and Salam shared the Nobel Prize in 1979 for this work.

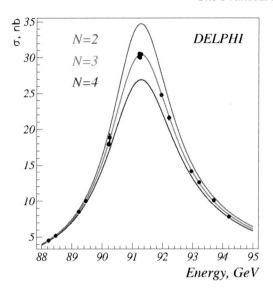

Fig. 5.3. The height of the curve measures the number of Z^0 particles that are made at the energy shown on the bottom axis. The peak corresponds to the mass equivalent of the Z^0. The width of the peak arises from the short lifetime of the Z^0, in accord with the Heisenberg uncertainty principle. Three calculations shown by the colored curves correspond to what the experiment would look like if there were two, three or four neutrino types. The data indicates unambiguously that there are three flavors. *Credit: CERN, Geneva.*

To make these particles now, in our time, it is necessary to provide in a very small space the energy equivalent of their great mass ($E = mc^2$). To achieve this, a five-mile-diameter collider was built at CERN, Geneva to provide a colliding beam of electrons and antielectrons (commonly called positrons), each having at least half the needed energy to produce the mass of the Z^0. However, it was found that a few Z^0s were created at slightly lower energy and some at higher energy, as shown in Figure 5.3. Peak production occurred at the mass of the Z^0. This type of variation in the energy that is needed to form a nuclear particle is well known in physics; it is exactly what is expected from the Heisenberg uncertainty principle if the particle does not live long, but decays into other particles. In that case the span in energy (ΔE) in which Z^0s are made, multiplied by the life (Δt) of the Z^0 particle, must be about the size of Planck's constant or bigger, according to Heisenberg. These words can be written as an inequality:

$$\Delta E \Delta t > h \,.$$

The width of the peak in the production of the Z^0 can be analyzed in terms of the various ways that the Z^0 can decay. Three curves are compared in Figure 5.3 with the data, assuming the number of types of neutrinos to be 2, 3, or 4. The answer is unambiguously 3, as can be seen from the figure.

However, this conclusion was not immediately evident. Surprisingly, the data varied throughout the day, shifting back and forth. Because of the twice-daily occurrence of this variation, it was soon recognized as being a tidal effect of the Moon's passage. The Moon was creating a small tide in the Earth's crust that changed the diameter of the accelerator in the direction of the Moon's crossing by about one millimeter out of a five-mile diameter: this caused the beam energy of the accelerator to vary by about one part in 10 000, which was what produced the two-peak pattern shown in Figure 5.4. When the data was corrected for this spurious effect, the smooth single peak shown in Figure 5.3 was obtained. It is an amazing testament to the accuracy with which the LEP collider and its instrumentation were designed and built, that such accuracy could be achieved.

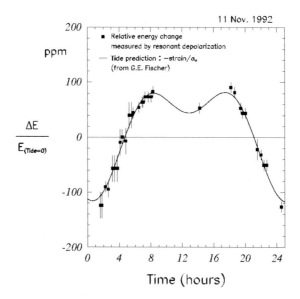

Fig. 5.4. This data that was collected over a 24-hour period shows the shift of the peak energy of the previous plot (Figure 5.3) caused by the Moon's tide acting on the accelerator and its mooring to the Earth. The tide on the Earth's surface stretched the diameter of the laboratory accelerator (which is buried 100 meters underground) by about one millimeter in its 5-mile diameter. *Credit: CERN, Geneva and the DELPHI project.*

The actual discovery of the third flavor had to await experiments performed at the Stanford Linear Accelerator, where the *tauon*, the heaviest version of the electron, was discovered in 1975 by M. Perl, who shared the 1995 Nobel Prize with F. Reines, the earlier discoverer of the first type of neutrino. The *tau* neutrino was finally discovered by a large international collaboration of scientists at the Fermi Laboratory (Chicago) in 2000. Thus, finally, laboratory experiments had confirmed an intimate property of the fundamental particles of nature that, for the first time in history, had previously been uncovered by the cosmological processes of helium production in the early universe. The measured abundance of helium provides a stringent test of the ideas presented in the discussion of this chapter. Let us see how it goes.

Above a temperature of $T = 10^{10}$ degrees, a supply of neutrons is produced by the (weak) interactions

$$e^- + p \longleftrightarrow \nu + n \,, \tag{5.1}$$

$$\bar{\nu} + p \longleftrightarrow e^+ + n \,. \tag{5.2}$$

Neutrons are an essential ingredient of atomic nuclei. Because they carry no charge, they contribute their strong nuclear force to binding neutrons and protons together in a nucleus while at the same time diluting the repulsion of the Coulomb force among the protons. So the two processes of neutron production are crucial to the formation of nuclei in the early universe. In the above reactions, e^+ denotes the positively charged antielectron, $\bar{\nu}$ the antineutrino, and $\longleftrightarrow$ means that all of the particles are in thermal equilibrium, and the reaction flows in either direction with equal facility. Notice that the first channel for production of a neutron requires an *electron* and the second an *antineutrino*. These two particle types will become so rare at some point in time that neutron production is cut off. When this happens the primordial synthesis of elements will soon end; what was made in those few minutes will be the entire supply of light elements in the universe to this very day. So the end of the first second is a very critical moment.

The neutron is more massive than the proton by an equivalent temperature of 1.5×10^{10} degrees and so normally will decay. But, as long as thermal equilibrium is maintained by the frequency of collisions, an equilibrium number of neutrons (N_n) as compared to the number of protons (N_p) will be maintained by the above reactions; the thermodynamic law of Boltzmann gives their relative number as

$$N_n/N_p = e^{-(m_n - m_p)c^2/kT} \,, \tag{5.3}$$

where k is Boltzmann's constant, T is the temperature, and m_n and m_p denote the neutron and proton mass. When T is very large the heat energy kT is much larger than the mass difference, the exponential factor is nearly e^{-0} (which is 1), and the number of neutrons is almost equal to that of protons. However, when T is small or zero, the ratio of mass difference to temperature is very large, and consequently the exponential is small or zero, in agreement with the assertion that neutrons will disappear ultimately as the temperature falls.

The energy equivalent of temperature $T = 10^{10}$ degrees or above is greater than the sum of the masses of the electron and positron (antielectron). Therefore, photons abound with energy sufficient to create electron and positron pairs by the reaction

$$\gamma \longleftrightarrow e^- + e^+. \tag{5.4}$$

The universe is swarming with electron–positron pairs so that the first of the pair of reactions, (1), provides a steady supply of neutrons. However, as the temperature falls below the equivalent energy needed to create pairs, electrons and positrons fall out of thermal equilibrium. Most electrons are lost then through pair annihilation (the above reaction takes place only in the reverse direction). At this point, electrons become so rare in the universe that the first of the two neutron sources, (1), is lost.

The other source of neutrons, (2), is lost when neutrinos and antineutrinos fall out of equilibrium and thereafter stream *freely* through the universe as if nothing else were there. How long before they decouple from the other contents of the universe, thus shutting off the creation of neutrons that are needed for the synthesis of nuclei? Recall that neutrinos interact very weakly. That is the reason they were so difficult to discover in the first place. Consequently, the frequency between interactions of neutrinos with matter will fall as the density of matter falls during the expansion. Indeed, the time between interactions will eventually become so long that it exceeds the age of the universe at every instant. After that, the primordial neutrinos will move freely forever, as if nothing else existed, as they do now.

The calculation of the moment at which neutrinos decouple is quite involved, so we only sketch it. The timescale for neutrino reactions will decrease with temperature to some power, because the density falls with decreasing temperature. We can figure out what that power is in the manner indicated in Box 13. We find that the time needed for neutrinos to interact with matter in a universe whose density is falling in the manner determined by the Friedmann–Lemaître equation becomes larger than the age of the universe at about $t = 1$ second (Box 14). The temperature of the universe at this stage is about 10^{10} degrees. Thereafter, neutrinos stream freely through the universe without interacting.

The loss of neutrinos from equilibrium closes the only other channel for neutron production. Neutrinos decouple at about the same time and temperature as electrons become scarce. Both particles are necessary for maintaining neutrons in thermal equilibrium with the rest of the universe. By the time that neutrinos move through the universe without again interacting with matter, the number of neutrons has fallen with the temperature to a value of about one neutron for every five protons ($N_n/N_p = 0.2$).

The above ratio becomes frozen from equilibrium at the moment neutrinos decouple. Neutrino decoupling places a strict upper limit on the number of light elements that can be made in the early universe because it takes one neutron for every proton to make a deuterium or helium nucleus. There is one additional barrier to element production in the early universe — a race for time. Not only is the supply of neutrons limited, but also they are unstable on their own and will spontaneously decay in about 10 minutes. That limits the period during which elements can be made in the early universe.

5.4.4 *Formation of the light elements (t = 100 seconds to 10 minutes)*

As long as the supply of neutrons lasts, they can combine protons under the attractive nuclear force to form deuterons.[17] The reaction is denoted by the equation

$$n + p \rightarrow D + \gamma.$$

The mass of the neutron and the proton together is greater than the mass of the deuteron. Therefore, they become *bound* together to form a deuterium nucleus, and the amount of energy carried off by the gamma ray is equal to the binding energy together with any kinetic energy of the nucleons. The neutron and the proton remain bound together unless some process occurs, such as the collision with a proton or a photon having an energy at least as large as the binding energy, in which case this amount of energy may be transferred to the neutron and the proton, thus liberating them.

---

[17]The idea that elements might be formed in the early universe dates from 1946 in the work of Gamow and his collaborators, Alpher, Follin, and Herman. On the heels of the discovery of the cosmic background radiation, Peebles wrote a code to follow the formation of helium-4; P.J.E. Peebles, *Astrophysical Journal*, Vol. 146 (1966), p. 542.

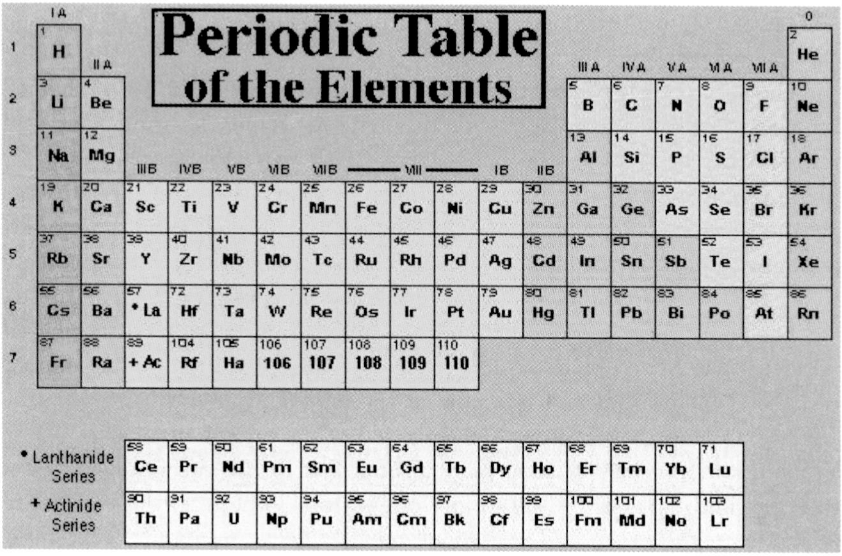

Fig. 5.5. Elements of the periodic table. Until about one-and-a-half minutes, the universe was too hot for any element to exist. Then, in the next three-and-a-half minutes, virtually all of the deuterium, helium, and lithium nuclei that exist today were formed (those with proton number up to 3, marked 1, 2, and 3 at the top of the table). Hundreds of millions of years later, after stars had formed, the heavier elements were fused in their cores, and in the gases that were expelled at the death of each star in a supernova explosion.

The equivalent temperature of deuteron binding is $T = 2.6 \times 10^{10}$ degrees. Temperature determines what the most *probable* energy of a photon will be, but in a substance in thermal equilibrium there are photons of both higher and lower energy than the prevailing temperature. The law that determines the number of photons of various energies when the temperature has a certain value was discovered by Planck, and in the reasoning that he used to arrive at it, he introduced for the first time the notion that radiation comes in quantized bundles of energy that we call photons (and gamma rays when the energy is exceedingly large). So at the temperature equivalent of the binding energy and even somewhat lower, there are still many high energy gamma rays that will destroy deuterons. This situation prevails until 100 seconds later, when the temperature has fallen to 10^9 degrees. Below that temperature, deuterons are stable and the gateway to production of other light elements has opened. At this stage, the neutron-to-proton ratio has fallen a little further, to $N_n/N_p = 1/7$.

In the next 200 seconds a chain of reactions that make other light elements consumes a large fraction of deuterons:

$$D + D \rightarrow {}^{3}He + n \,,$$

$$D + D \rightarrow {}^{3}H + p \,.$$

Now the deuterons (D) produced in the first fusion reaction combine with the tritium $({}^{3}H)$ from the second to form ${}^{4}He$:

$$ {}^{3}H + D \rightarrow {}^{4}He + n \,.$$

Because of their strong binding, essentially every ${}^{4}He$ that was formed survived. If the entire supply of neutrons went into helium production at this point, the fraction of baryon mass in the universe that would reside in ${}^{4}He$ would be 25%.[18] This figure is remarkably close to the actual measured abundance.

The supply of neutrons was being eroded as these light elements were formed. It was now only a matter of time before the remaining few minutes of the neutron half-life caused any remaining unbound neutrons to disappear (Figure 5.6). Detailed calculations have shown that in the succeeding few minutes all neutrons either decayed or were incorporated in the synthesis of deuterium and helium to form the primeval abundance of these elements that exist to this day.[19]

The large binding energy of helium-4 $({}^{4}He)$ ensures that it is relatively insensitive to the baryon density. Not so for deuterium $({}^{2}H)$. The abundance of deuterium in the universe is a key to verifying the theory of nucleosynthesis in the first few minutes. Deuterium is the gateway to production of the primordial elements $({}^{3}He, {}^{3}H, {}^{4}He, {}^{6}Li \ldots)$ The reaction rates that produce these elements occur faster the higher the baryon density. And the deuteron abundance is a steeply falling function of baryon number as it is consumed in their making.

The density of *high* energy photons also erodes the abundance of deuterium by disintegrating them. The larger the *ratio* of baryons to photons,

---

[18]Because it takes two neutrons to make helium-4, the ratio of that element made in the early universe to all nucleons is $2N_n/(N_n + N_p)$, which is 1/4 when $N_n/N_p = 1/7$.

[19]Detailed calculations of nucleosynthesis in the early universe were published in two classic papers: R.V. Wagoner, W.A. Fowler, and F. Hoyle, *Astrophysics Journal*, Vol. 148 (1967), p. 3; R.V. Wagoner, *Astrophysics Journal*, Vol. 179 (1973), p. 343. Wagoner's computer code has been updated from time to time as more accurate nuclear data becomes available.

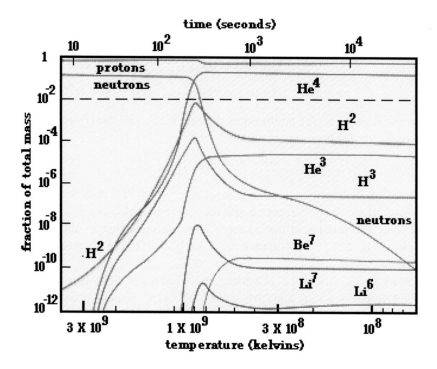

Fig. 5.6. The *calculated* light element abundances as they change with time during the first few hundred seconds in the life of the universe. Before 100 seconds, almost no deuterium (denoted on the graph by H^2) had been made. By 300 seconds the cosmic supply of helium had been made. Aside from those elements that undergo radioactive decay, the abundances remain fixed after a few minutes up to the present day, when they have been measured. *Permission: L.F. Thompson and F.H. Combley.*

the smaller the production of deuterium. These trends are illustrated in the *computed* abundances shown in Figure 5.7. The vertical shaded region indicates the range of baryon density in which, within observational error, the observed abundances *all* agree. In this way we learn that the Big Bang model of primordial element synthesis during the first 10 minutes agrees very well with observation. Moreover, because these computed abundances depend on the assumed number of baryons per photon at the time of synthesis, it was found that there are an enormous number photons compared to baryons:

$$n_\gamma / n_B = 2 \times 10^9 \,.$$

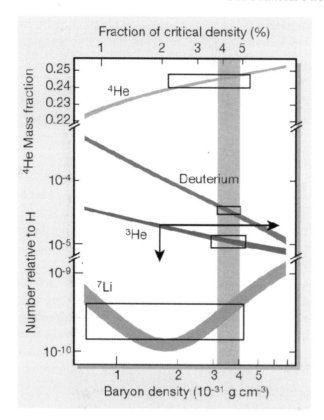

Fig. 5.7. The abundance of helium-3, helium-4, deuterium, and lithium-7 in the universe plotted as a function of the *present* baryon density. The blue bar marks the range of the most sensitive measure, deuterium, which also agrees with the abundances of the other isotopes. This data indicates that the best estimate of the present matter density is $\rho_B = (3.5 \pm 0.4) \times 10^{-31}$ g/cm^3. *Permission: C. Charbonnel and* Nature.

But this ratio has not changed in all the intervening billions of years because baryon number is conserved, and the photon number also has remained undisturbed since the decoupling of radiation and matter at 300 000 years (Section 5.4.6). Between element production and that time it did not change much either because the number of photons is overwhelmingly larger than the baryon number, so that relatively few photons could have been absorbed or scattered. These facts allows us to calculate the *present* baryon mass density. This is how.

Using the discovery by Penzias and Wilson of the cosmic background radiation temperature, now measured accurately as $T_0 = 2.728$ degrees

Kelvin, we know how to calculate the present photon density from Planck's law for blackbody radiation (Box 16). It is

$$n_\gamma = 413 \text{ photons/cm}^3 .$$

Hence, since we know the baryon-to-photon ratio from its determination of the deuterium abundance (see above), we can find immediately the *present* baryon mass density. It is

$$\rho_B = 3.5 \times 10^{-31} \text{ grams/cm}^3 .$$

For future reference we note that the baryon density deduced from primordial nuclear synthesis represents the present baryon density *whether or not* it is visible to us now in stars and galaxies. What cannot be seen is called *dark* baryon matter, simply because it is not seen. Its hiding places may be in defunct dim stars, neutron stars, and white dwarfs.

Let us describe the significance of the above value of baryon density in another way: if all the planets, stars, and galaxies that are visible together with all unseen baryons whether in failed stars, brown dwarfs, or whatever, were smoothly spread throughout space, then on average there would be one baryon in every 5 cubic meters of space. Space is almost empty of baryons, while at the same time baryons account for almost all the mass of ordinary matter in the universe.

5.4.5   *Measurement of primordial abundances of elements*

Helium is found everywhere in the universe, in the atmospheres of old stars and young alike, in our galaxy and others, and in very distant quasars. Other elements are also found, but their abundances vary strongly according to the bodies they are detected in. Not so for helium. Everywhere that helium is detected there is about one helium for every ten hydrogen nuclei, which by mass is about 25%, just as predicted by the Big Bang theory of nucleosynthesis. How can the helium, or any other element, be detected in the atmosphere of some far-off star? Of course, by the line spectra emitted or absorbed by ionized atoms of the elements, which produce unique patterns by which their presence can be identified. (Recall the discussion of line spectra on the pages starting at 22 and also the figure on page 93.)

However, there is one matter that still has to be resolved. Have the light element abundances, which can be measured only in our time, remained unchanged through these 15 billion years since the first few minutes? In other words, are there processes by which these elements could have been made or destroyed subsequently? If so, they would hide from us the original abundances. Do these great furnaces, the stars, manufacture enough of the light elements that the proposed means of determining the *primordial* abundances and hence the nucleon-to-photon number ratio would be spoiled?

The answer is no and here is why. All stars produce helium, stars as small as our Sun, or as large as Betelgeuse, the great star in the constellation Orion that has a radius 630 times that of the Sun. But even those not much more massive than our Sun go further — from the helium they produce carbon — then much of the carbon with some remaining helium produces oxygen. Stars having a mass like our Sun burn very slowly. The Sun is 4.5 billion years old and will live to be 12 billion years before expelling much of its mass and sinking to the white dwarf stage. In the universe, with an age of 15 billion years, our Sun is only the third or fourth generation of small stars; consequently, small stars cannot account for very much helium in the universe. What of heavier stars?

Stars with masses 10 times that of the Sun go on to synthesize elements all the way to iron and nickel. In other words, heavier stars convert most of the helium that is produced in their early years to still much heavier elements in their later years. Such stars live only about 20 million years before they explode, thereupon releasing about 10 times as much energy as they did during their entire prior lifetime as luminous stars. When these stars explode, they distribute into the universe heavier elements that were not made in the first 10 minutes because the early universe cooled and diluted too rapidly. The gases of generation upon generation of exploded stars wandered the universe for eons, mixing with the gases of other explosions, forming yet new generations of stars that were enriched in these elements. Eventually, the concentration of the heavier elements became sufficiently high that solid planets, besides new stars, could form. And on at least one such planet around one insignificant star, plants, and animals like ourselves, arose by processes that we do not as yet know. But of the many generations of stars that have lived and died, so little helium survived that the present abundance cannot be accounted for because the stars burn it after making it (Box 18). The abundance of helium in the universe, about 25% by mass, is primeval.

As for deuterium, its abundance is very low — about one in 30 000 hydrogen atoms. It is hard to measure, and efforts to refine the measurements continue. Because of its very small binding energy, it is easily destroyed, and cannot survive in stars, so what is seen now must have been made in the first few minutes. Indeed, what exists now places an upper limit on the primordial production of deuterium.

5.4.6   *Decoupling of radiation and matter (t = 300 000 to 1 million years)*

For the first 300 000 years the universe was shrouded in a fog so dense that it was totally opaque. Until the universe cooled below 3 000 Kelvin, radiation and matter were in such strong interaction that light could not travel freely; photons could barely move before being interrupted by a collision with electrons and sent off in another direction, only to be interrupted again, and again.... They were in thermal equilibrium with matter, and moved with the flow of matter as the universe expanded. We have learned of the vastness of the cosmos now, so that when astronomers look deep into space with powerful telescopes they are looking also backward in time. That is because of light travel time. But it is impossible to "look" further back into the time when radiation and matter were strongly coupled because light was trapped by the particles that it frequently collided with in that high density environment.

The situation is quite analogous to light on a cloudy day. We look up and see the undersurface of the clouds. They are bright with light and we know that the Sun shines above, but we cannot see it. The only light we receive is from the bottom thin layer of the clouds. Light from the Sun has been reflected back and forth between water molecules in the cloud, and only the light that reaches the bottom after many scatterings can shine down. That bottom surface is the *surface of last scattering*. Likewise, looking into the distance with powerful telescopes, astronomers can see only as far back as the decoupling era. All that happened before can be learned only through inference, such as the abundance of light elements that were formed prior to decoupling, the cosmic background radiation, and, perhaps someday, the neutrino background.

As we learned above, we do have evidence of what took place at those early times. The measured abundance of helium constrains very narrowly the scenario that we have just discussed in the previous sections. If it were seriously in error in any of its many details, the calculation of the abundance would not have agreed with observation. We have seen that it does. But we would like further evidence that the description of how the early universes evolved is really correct.

We have already described another piece of evidence — the cosmic background radiation which Penzias and Wilson discovered. It is the messenger from that instant when the fog lifted and the light shone forth. By now it has been measured in all its fluctuating detail, so that the very *seeds* of galaxies have been detected (Figure 3.3). That part for much later (Section 7.3.2).

First let us look at the details of decoupling to see how accurately we can determine the time at which the universe became transparent.

Recall that the point in time when we left off was the era when the light nuclei were made. At the end of that time, the universe was still filled uniformly with radiation and with electrons, protons, and those few nuclear types — deuterons, helium-3, helium-4, lithium-7 — that were made in the first few minutes. Nuclei are composed of neutrons and positively charged protons that normally attract one electron for each proton: the electrons form a halo around the nucleus, the whole constituting a charge-neutral *atom*.

But, the universe was much too hot when nuclei were first formed for them to collect electrons: the random jostling associated with heat was much too vigorous for electrons to stick to nuclei. So the universe was still filled with a plasma of photons interacting with charged particles. And any electron that did become bound to a proton or a nucleus would shortly be knocked out again. Particles and photons were so closely packed and the temperature so high that the time between collisions among themselves was very short compared to the time it took for the density to change appreciably due to the universal expansion. The entire contents of the universe, at this early time, were in adiabatic thermal equilibrium.

Gradually, however, as the temperature dropped, one electron would attach itself to a nucleus, and remain — then a second one — but because the first electron neutralizes the charge of one proton, the attraction of the nucleus is not as strong as before. So the second electron will suffer the fate that the first had earlier suffered. However, with the temperature ever dropping, one after another additional electron would become attached and remain attached to a nucleus.

The critical moment we are looking for is when, one by one, all electrons became bound in nuclei to form charge neutral atoms. This occurred as the temperature dropped below about

$$T_{\text{decoupling}} \approx 3000 \text{ K}.$$

Meanwhile, as the universe expanded and cooled, the wavelength of the radiation become so long through the Doppler effect that the photons could not distinguish the electrons in the atom from the protons in the nucleus; so they felt no charge at all. At that moment the flight of photons at the speed of light was uninterrupted, and they streamed freely through the expanding universe even to this day, just like the neutrinos before them. Those photons constitute the cosmic background radiation that permeates space. It remains unchanged except for the Doppler shift in wavelength caused by the expansion of the universe. When radiation is in thermal equilibrium with matter, meaning that radiation and matter have the same temperature, the relative number of photons at each wavelength depends only on the temperature. This is Planck's law. When radiation became decoupled from matter, the two no longer kept the same temperature. But, each photon wavelength was stretched by the expansion of the universe ($\lambda \sim 1/R$), so its energy decreases in proportion to the inverse of the expansion scale. The Planck law remains unchanged when all photon energies change in this way; the temperature of radiation simply decreases as $T \sim 1/R$, just as it did when it was interacting with matter.

5.5   Close of the Radiation Era

From the time all electrons became bound to nuclei to form charge-neutral atoms, the radiation of the early universe and matter no longer interacted with each other except gravitationally. This remaining interaction was important in the early phase of galaxy formation. However, as the universe continued its expansion, the density of radiation diluted more rapidly than that of matter. The cause of this difference is the Doppler shift of radiation, as we will understand shortly (page 149). In the late stage of the radiation era, matter, which up to that time had been uniformly spread through the cosmos, began to collect under the attraction of gravity into enormous clouds of hydrogen and helium with traces of deuterium and lithium.

Cloud formation and collapse to form galaxies and galaxy clusters became more pronounced as the density of radiation faded in importance. A new era was beginning: the matter era — the era in which galaxies were formed and stars were born, matured, and died. This is the era when heavy elements, so essential for life, were manufactured in maturing stars and sent forth by great explosions called *supernovae* into the universe, there to wander for eons. These elements mixed with the refuse from other explosions in the diffuse gas of great clouds. Eventually parts of the clouds became unstable and collapsed under their own gravity to form new generations of stars (Figure 5.8). And after many generations of stars, the great clouds of gas contained enough of these heavier elements that when they began their collapse to form new stars, pieces broke away to form solid planets. Let us begin now to study this phase of structure formation in the universe.

Two important events occurred near the same temperature, but they are distinct. We have just discussed decoupling of radiation and matter at 3000 K. A little later, at about 2000 K, the density of radiation fell below that of matter. Matter became dominant after that.

Fig. 5.8. In our own galaxy, this beautiful pillar of cool molecular hydrogen gas and dust is a nursery for newborn stars. It is called the Eagle Nebula (M16) and is located in the constellation Serpens at a distance of 6500 light years. Baby stars are embedded inside finger-like protrusions called EGGs (evaporating gaseous globules), extending from the top of the nebula. The EGGs, each somewhat larger than our solar system, are brightened, heated, and in the process of evaporation by the ultraviolet light from nearby hot stars. The growth of the baby stars, which are gathering gas from the EGGs, will cease when they are uncovered by the evaporation of the EGGs. They will be the next generation of stars in our galaxy. *Credit: NASA, J. Hester and P. Scowen, Arizona State University.*

5.6   Matter Dominance

Let us imagine a very large sphere, R, that is moving *with* the flow of the expansion. In this case, nothing enters or leaves the sphere, neither in the past nor in the future. Light travels faster than any matter can, so we define comoving spheres in terms of their material content. This is a reasonable approximation during the later life of the universe because radiation now accounts for so little energy compared to matter. It is also a good approximation for the very early universe: the density of charged particles was so large that radiation could move only the distance separating several particles before it was scattered and changed direction and possibly shared its energy. The universe was opaque at early times. Light did not travel freely but moved with the flow of matter.

Because of the universal homogeneity and isotropy, any such sphere as R is equivalent to any other. So we can restrict our discussion to what happens inside R. Whatever happens in R happens everywhere. R is called the scale factor and was discussed in Section 3.6. Einstein's theory determines how it behaves with time. For the present, we are interested only in the general conditions of the contents of R at earlier times of the universe. We place a time label on $R(t)$ to distinguish its value at different times. We measure time from some distant past, so we call it *cosmic time*. We denote the present time as t_0.

In the cosmic era in which we live, the mass density of radiation, i.e. of photons, is much less than that of matter, and both are very small.[20] Also in the present universe, there are very few free electrical charges. Almost all electrons reside in neutral hydrogen and helium atoms, and in heavier atoms, in which the number of electrons, which have a negative unit of charge, and of protons, which have a positive unit, are equal. Photons interact with electric charge but not with neutral atoms, unless the photons are of very high energy (short wavelength), which is not the case in the present epoch. Consequently, as concerns the large scale or average behavior of radiation and matter, it is as if the other did not exist. They are said to be uncoupled or decoupled.

The number of nucleons is a conserved quantity. It obeys one of the important conservation laws that we discussed in Chapter 4. (The half-life of the proton, if it decays at all, is determined to be greater than 10^{33} years, which is a billion times the age of the universe.) And the photons that came out of the early fireball at a time when the universe was only 300 000 years old have not interacted since then, so their number has remained unchanged.[21]

The density of nucleons and photons is the number of them in a cubic centimeter, so the densities do change as we look backward or forward in time. Since by our definition the boundary of the sphere R moves with the flow of matter, the matter density changes inversely to the change in volume; that is to say, matter density varies as $1/R^3(t)$. The number density of photons also varies in this way even though their speed is c and

---

[20]The average number of nucleons in the universe is 1 in every 5 cubic meters.

[21]When the universe had sufficiently cooled, all electrons became bound to protons and the other very light nuclei. At that moment, radiation (which can interact only with electrical charges) and matter ceased to interact. This is referred to as the *decoupling* of radiation and matter, and the coupling of electrons with protons and nuclei to form charge-neutral atoms is referred to as the combination era (sometimes the recombination era).

therefore they move faster than the surface of the comoving volume; but as many enter as exit because of the uniformity and isotropy of the universe. However, unlike matter, the number of photons does not tell us their energy: their wavelengths are stretched in an expanding universe, and therefore the energy carried by each photon decreases with time.[22] In fact it decreases as the universe expands in proportion to $1/R(t)$. Consequently, the equivalent mass density of radiation varies as $1/R^4(t)$.

We can draw a very important conclusion from the fact that the equivalent mass density of radiation and of matter scale differently:

$$\rho_r/\rho_m \sim 1/R.$$

(The subscript r denotes radiation, and the subscript m matter.) We conclude that because the mass density of radiation is less than that of matter *now*, it must have been equal at some distant time in the past, and greater at all preceding times. We refer to early times as the *radiation-dominated era*. At some distant time in the past, a time we will derive later, the universe changed from radiation-dominated to matter-dominated; that is the era we are in at present. The universe will remain in this era as long as the cosmic expansion continues. In the radiation-dominated era, the light elements of the periodic table were formed (Figure 5.5). In the matter-dominated era, galaxies, then stars, and then heavy elements were formed, as we will discuss in the following chapters. The actual conditions under which the universe switched from being radiation- to matter-dominated are therefore of great importance concerning the abundance of light elements like hydrogen and helium, and concerning the formation of galaxies and galaxy clusters. Let us find out that time and temperature.

The total baryonic mass density, both luminous and dark, is presently $\rho_B \approx 3.5 \times 10^{-31}$ g/cm^3. We learned this in our study of primordial nucleosynthesis (page 142). However, from the work of many astronomers over the years, the approximate number of galaxies is known and also their masses. From that assay, the density of *luminous* baryonic matter in galaxies in the form of stars and dust is $\rho_G = 3 \times 10^{-32}$ g/cm^3. Therefore, only about $1/10$ of baryonic matter is luminous. Most baryons are dark; that is to say, they are contained in nonluminous bodies and gas that are not visible to us. Only a fraction of gas in interstellar space is visible to us, being illuminated by nearby stars, as in Figure 5.8.

---

[22]The Doppler shift changes the wavelength according to $\lambda = [R(t)/R(t_0)]\,\lambda_0$ (page 22). Consequently the photon energy changes according to $hc/\lambda \sim 1/R(t)$.

As to the density of radiation, it has been known since the early part of the 20th century how to calculate the equivalent mass density of radiation when its temperature is known. The law that provides the answer is the Stefan–Boltzmann law, which tells us that the equivalent mass density of radiation is proportional to the fourth power of the temperature, i.e. $\rho_r = aT^4/c^2$. It also provides the value of the constant of proportionality, a. From the present temperature of the relic cosmic background radiation, 2.728 K, we calculate the present equivalent mass density of radiation to be $\rho_r \approx 4.7 \times 10^{-34}$ g/cm^3 (see Box 16). From the above two densities we find that the ratio of baryonic mass density (visible and dark) to radiation is approximately 750. From the difference in the way that radiation and mass densities scale, we know that the universe has expanded by the same factor between the present time and that long-ago time when matter and radiation densities were equal (Box 17).

What was the temperature at that earlier time? The key, again, is the Doppler shift of radiation. The wavelength of all radiation stretches in proportion to the universal expansion, and the temperature therefore falls in inverse proportion to the expansion. Consequently, when radiation and matter were equal in mass density the temperature must have been larger by the factor by which the universe has since expanded:

$$T_E = 2.7 \times 750 = 2000 \text{ K}.$$

Fritz Zwicky, already in the 1930s, discussed how the observed rotation of galaxies and also the motion of galaxies in clusters tell us that there is much more matter in and around galaxies than is contained in the luminous stars and dust that can be seen. The stuff of stars and dust we call baryonic matter. It is in the form of atoms, and by far the greater part of their mass is contributed by baryons — the protons and neutrons. The matter which cannot be seen, but which is inferred to exist in galaxies, and even between galaxies in clusters, is called *dark matter* and, as we will also learn, most of the dark matter is *nonbaryonic*. Its precise nature is not known. All matter, luminous and dark, experiences the universal force of gravity. Otherwise, nonbaryonic dark matter interacts very weakly with ordinary baryonic matter made up of protons and neutrons. The latter interact through the strong nuclear force. But, nonbaryonic dark matter is presumed to consist of particles of an unknown type that do not experience the nuclear force. Not even the most sensitive laboratory experiments devised so far have been able to detect them.

The temperature of the universe at the time when radiation and matter had equal mass densities marks an important transition point concerning the formation of galaxies from the featureless universe that we have discussed so far. We can estimate how old the universe was at that time of transition from the dominance of radiation to the dominance of matter. Recall that the Friedmann–Lemaître equation on (page 74) governs how the scale of the universe changes with time. In the radiation-dominated era, that differential equation takes on an especially simple form (see Box 10 for mathematical details). Readers with a mathematical bent will be able to confirm with a few manipulations that the solution to the Friedmann–Lemaître equation for the expansion of the universe is equivalent to an equation for the temperature T whose solution as a function of cosmic time t can be written down as $T^2 = \sqrt{3c^2/32\pi Ga}\,/t$. When the various universal constants, G, c, a, π, are inserted we find that the universe was about a million years old at the time its temperature had fallen to 2000 K, the time when radiation and all types of matter — luminous and dark — had equal mass densities (Box 11):

$$t_E \approx 10^6 \text{ years when } \rho_r = \rho_B\,.$$

The universe was less than one 10 000th of its present age when matter began to dominate radiation density.

Is it not a wonder that using a few of the laws of physics and given two pieces of data — namely the approximate present value of the matter density in the universe and the measurement by Penzias and Wilson of the present temperature of the background radiation — we have been able to know the age of the universe and its temperature at such an early fiery time?

Of course, the real fireworks began much earlier. After all, 2000 K is not a very high temperature. Most metals melt at that temperature and the Sun's surface is 5000 K. But before looking further toward our own time, let us look still further back to the earliest time and trace the cosmic events that have taken place since then to make the universe we live in now.

5.7 Boxes 9–20

9 Temperature Variation with Expansion

The wavelength of radiation λ is stretched by the universal expansion according to

$$\lambda_0/\lambda = R_0/R\,,$$

just as the wavelengths of photons in a box are stretched if the box dimensions are increased. The number of photons is conserved because the number of photons is 2×10^9 (Box 16) times the number of baryons, so that photon scattering is extremely rare. (It becomes impossible, except for photon–photon scattering, after electrons have combined with protons and nuclei, referred to as decoupling. This event is sometimes inappropriately referred to as *recombination*.) It follows that $N \sim 1/R^3$, so that the energy density

$$\epsilon(\nu) = \sum h\nu N(\nu)$$

transforms as

$$\epsilon/\epsilon_0 = (R_0/R)^4\,.$$

The spectral distribution depends on $h\nu/T$, so that T is altered in the same way as ν, namely

$$T/T_0 = R_0/R\,.$$

10   Evolution of Early Universe

Of the three terms on the right side of the Friedmann–Lemaître equation (page 74), only the one containing the density of mass, ρ, is important at early times. Therefore the universal expansion is governed by

$$\dot{R}^2 = [8\pi G\rho(t)/3]\, R^2(t)\,.$$

Radiation dominates in the early universe, so from the Stefan–Boltzmann the equivalent mass density is

$$\rho(t) = aT^4(t)/c^2\,,$$

where k is the Boltzmann constant, and

$$a = 8\pi^5 k^4/15(ch)^3$$
$$= 7.57 \times 10^{-15} \text{ g}/(\text{cm s}^2 \text{ K}^4)\,.$$

The wavelength of radiation is Doppler-shifted by the expansion so that $T(t) \sim 1/R(t)$, and consequently

$$\dot{R} \sim -\dot{T}/T^2\,.$$

Therefore the Friedmann–Lemaître expansion equation can be written as a time evolution equation for the temperature:

$$\dot{T} = -\sqrt{8\pi Ga/3c^2}\, T^3(t)\,.$$

The solution is
$$T^2 = \sqrt{3c^2/32\pi Ga}\,/\,t\,.$$

Therefore, in the early universe, temperature decreases as $T \sim 1/\sqrt{t}$, density decreases as $\rho_{\text{radiation}} \sim 1/t^2$, and the universe expands as $R \sim \sqrt{t}$. In particular

$$R(t) = T_0 R_0 (32\pi Ga/3c^2)^{1/4}\sqrt{t}\,,$$

where T_0 is the present temperature of the background radiation and R_0 is the present scale factor, which can be taken as unity. Inserting the fundamental constants, we find for the factor appearing in the above equation

$$\sqrt{3c^2/32\pi Ga} = 2.31 \times 10^{20} \text{ s K}^2\,.$$

For the density we find

$$\rho = (3/32\pi)(1/Gt^2)\,.$$

In summary, the mass equivalent of the radiation density scales as

$$\rho \sim T^4 \sim 1/R^4 \sim 1/Gt^2\,.$$

11   Temperature and Density of the Early Universe

The results of Box 10 took account only of photons. It is trivial to improve those results. At high temperature when kT is large compared to particle masses, the equilibrium number of particles (N) and antiparticles ($\bar{N}$) in a vacuum or the early universe is given by statistical mechanics as

$$N = \bar{N} = 4\pi g/h^3 \int_0^\infty p^2\, dp/(e^{pc/(kT)} \pm 1)\,,$$

where g is the statistical weight of the species, the $+$ sign holds for fermions and the $-$ sign holds for bosons. The following results apply.

Photons and bosons:

$$N = 0.488\, x^3 \text{ meter}^{-3}\,, \quad \epsilon = aT^4\,.$$

Electrons (each flavor), nucleons, hyperons, and their antiparticles:

$$N = \bar{N} = 0.183\, gx^3 \text{ meter}^{-3}\,, \quad \epsilon = (7/8)g\, aT^4\,.$$

Neutrinos and antineutrinos (each flavor):

$$N = \bar{N} = 0.091\, x^3 \text{ meter}^{-3}\,, \quad \epsilon = (7/16)g\, aT^4\,,$$

where $x = 2\pi kT/hc$, and a is the Stefan–Boltzmann constant. The total energy density therefore has the form

$$\epsilon = \alpha(T)aT^4\,.$$

To take account of all these species, instead of merely photons, the results of Box 10 should be modified by the substitution $a \to a\alpha(T)$. For example, at an epoch during which the temperature is greater than the electron mass but less than the muon mass, i.e. when $10^{12} > T > 5 \times 10^9$ and there are photons, electrons, positrons, and three flavors of neutrinos and their antineutrinos, the degeneracy is

$$\alpha = 1 + 2 \times 7/8 + 2N_f \times 7/16 = 43/8\,.$$

However, when $T < m_e = 5 \times 10^9$ degrees, the degeneracy factor became $\alpha = 29/8$ (photons, and three flavors of neutrino pairs). Using the results of Box 10 we obtain the convenient connection between time and temperature in the radiation-dominated universe:

$$T = 1.5 \times 10^{10}/\alpha^{1/4}\sqrt{t}\,,$$

where T is in K and t is in seconds. Another useful relation gives the mass density of radiation:

$$\rho = aT^4/c^2 = 4.5 \times 10^5/\alpha t^2 \text{ g/cm}^3\,.$$

12   Planck Time

The condition that the de Broglie wavelength of the visible universe fits within the cosmic horizon defines the time we seek:

$$h/p = \lambda = ct\,.$$

In a volume of such a dimension, the number density of particles and photons behaves as

$$n \sim 1/\lambda^3\,.$$

In this confined region, particles are ultrarelativistic, so that their energy is

$$E \sim pc = hc/\lambda\,.$$

Consequently the equivalent mass density is

$$\rho \sim nE/c^2 = h/\lambda^4 c = h/c^5 t^4\,.$$

From Box 10 we have derived the mass density of radiation (keeping only dimensioned quantities) as

$$\rho \sim 1/Gt^2\,.$$

Combining the last two equations we find that

$$t \sim \sqrt{hG/c^5} = 5 \times 10^{-44} \text{ s}\,,$$

which is the time we sought, the Planck time, when the visible universe lay within its de Broglie wavelength.

13   Timescale of Neutrino Interactions

Neutrinos decouple from the rest of the universe when the mean time between interactions with matter exceeds the age of the universe. To calculate the age of the universe at that time, we first compute the interval between interactions.

The density of neutrinos varies as

$$N \sim 1/R^3 \sim T^3 \,,$$

because $T \sim 1/R$, where R is the scale factor of the universe. The cross-section for the weak neutrino reactions is

$$\sigma \sim G_A^2 (kT)^2 \,,$$

so the timescale for the reactions is

$$\tau \sim 1/\sigma N c \sim 1/T^5 \,,$$

where G_A is the weakinteraction coupling constant. By contrast, the scale factor varies as $R \sim 1/T$, so that the time between reactions very soon becomes longer than the age of the universe. Neutrinos drop out of equilibrium at one second.

14   The Neutrino Reaction Timescale Becomes Longer Than the Age of the Universe

We can calculate the time after which the interval between neutrino reactions (see Box 13) became longer than the age of the universe. Both the neutrino reaction timescale and that of the universal expansion are functions of temperature (Box 10). By eliminating temperature we find immediately

$$t \sim G_A^{4/3} (3c^2/32\pi G\alpha a)^{5/6} \,.$$

After this time the reaction rate is smaller than the expansion rate of the universe, so there are no neutrino reactions at all. We find that the frequency of neutrino interactions with matter ceased to be significant after $t = 1$ second.

15 Ionization of Hydrogen

The ionization energy of hydrogen is 13.6 eV. From the conversion

$$kT = 8.63 \times 10^{-5} \text{ eV},$$

this energy corresponds to a temperature of 1.5×10^5 K. Yet, at much lower temperature, hydrogen was ionized. The point is that the tail of the Planck distribution, *because* of the very high ratio of photons to baryons (2×10^9) from Box 16, contained a large number of ionizing photons.

16 Boltzmann and Planck Laws

From the present temperature of the radiation background, $T_0 = 2.728$ K, the *present* mass density of radiation can be calculated from Boltzmann's law:

$$\rho_r(t_0) = aT_0^4/c^2 = 4.66 \times 10^{-34} \text{ g/cm}^3.$$

The present *number* density of photons (i.e. the number per cubic centimeter) can be found from Planck's law for blackbody radiation and the measured cosmic background temperature. The number density is

$$n_\gamma(t_0) = 0.244(2\pi kT_0/hc)^3 = 413 \text{ photons/cm}^3.$$

From the baryon-to-photon ratio $n_B/n_\gamma = 5 \times 10^{-10}$ determined from primordial abundances (see page 140), we can calculate the *present* baryon *number* density as

$$n_B = 2 \times 10^{-7}/\text{cm}^3$$

and the mass density is then found to be

$$\rho_B = n_B m_N = 3.5 \times 10^{-31} \text{ g/cm}^3.$$

As an added note we can emphasize that the number of photons vastly outnumbers the number of baryons:

$$n_\gamma = 2 \times 10^9 n_B.$$

As a result of this, photons effectively *did not* encounter baryonic matter or electrons from a very early time in the history of the universe.

17 Expansion Since Equality of Radiation and Mass

From primordial nucleosynthes (page 142), $\rho_B \approx 3.5 \times 10^{-31}$ g/cm^3. The density of radiation can be found from the Stefan–Boltzmann law, $\rho_r = aT^4/c^2$. From the present temperature of the relic cosmic background radiation, 2.728 K, we find $\rho_r \approx 4.7 \times 10^{-34}$ g/cm^3 (Box 16). From the above two densities the ratio of baryonic mass density (visible and dark) to radiation is

$$\rho_B(t_0)/\rho_r(t_0) \approx 750.$$

The cosmic background radiation (CBR) is essentially the total mass density carried by radiation. The stars contribute only

$$\rho_*/\rho_B \approx 3 \times 10^{-5}.$$

By how much has the universe expanded since the radiation and baryon mass densities were equal? (Radiation photons do not have mass. But because of the Einstein equivalence of energy and mass, $E = mc^2$, a photon of frequency ν or (equivalently) wavelength λ has a mass equivalent of $h\nu/c^2 = h/c\lambda$.) Matter density and radiation scale with the expansion differently, as already noted. So the following two relations can be written:

$$1/R(t_0) \sim \rho_r(t_0)/\rho_B(t_0),$$

where t_0 denotes the present cosmic time, and

$$1/R(t_E) \sim \rho_r(t_E)/\rho_B(t_E) \equiv 1,$$

where t_E denotes the time at which the mass density of all baryonic matter and radiation were equal. Dividing the second by the first, it follows that

$$R(t_0)/R(t_E) = \rho_B(t_0)/\rho_r(t_0) \approx 750.$$

Thus the universe has expanded by 750, nearly a thousand-fold, since matter first became dominant.

The wavelength of all radiation stretches in proportion to the universal expansion, and the temperature therefore falls with expansion:

$$\lambda \sim R \quad \text{and} \quad T \sim 1/R.$$

So, at that earlier time when radiation and matter were equal in mass density, the temperature must have been larger than the present 2.7°:

$$T_E = 2.7 \times 750 = 2000 \text{ K}.$$

18   Helium Abundance

Stars, even though they synthesize helium, cannot account for much of the helium in the universe because they burn it to produce heavier elements. Or, aside from a little helium that is expelled into space by a wind from the flaming surface of low mass stars like our Sun, the helium is locked within them for 12 billion years or more (the expected lifetime of our Sun). Supposing that all the helium-4 observed in the universe were actually made in stars, the equivalent mass density would have to be greater than 0.002 times the density of mass in the galaxies (see next box). But, as we learned on page 158, the actual ratio of all starlight to matter is not greater than 0.00003.

19   Helium Abundance Is Primeval

It can be made quite clear that only a small fraction of the cosmic abundance of helium-4 can be synthesized in stars. Each formation of a ^4He nucleus releases a binding energy of 27 MeV while the mass of a ^4He nucleus is 3728 MeV. Almost all of the binding energy makes the light by which stars shine. So the radiation produced per ^4He is

$$\rho_*/\rho_{He} = 27/3728 \,.$$

From the observed abundance of ^4He,

$$\rho_{He} \sim (1/4)\rho_m \,,$$

where ρ_m is the estimated mass density of matter in the universe as determined by weighing and counting galaxies. In that case

$$\rho_*/\rho_m = (1/4)(27/3728) \approx 2 \times 10^{-3}$$

is the amount of radiation in starlight that we *should* see if all helium present in the universe today had been made in stars. But on page 158 we found that the ratio of the equivalent mass density of starlight to the density of matter is only 3×10^{-5}. So we must conclude that *if* the observed helium abundance were actually produced in stars, starlight would be $(2 \times 10^{-3})/(3 \times 10^{-5}) \approx 66$ times brighter than it is. Consequently, only a small fraction, if any, of the observed helium abundance could have been made in stars.

20   Redshift and Scale Factor Relationship

We seek the relation between redshift of radiation and the cosmic scale factor. Consider a faraway galaxy from which we receive light. One crest arrives at t_0 and the next at $t_0 + \Delta t_0$ and they were emitted at t_e and $t_e + \Delta t_e$ respectively. The light has traveled radially toward us on a null geodesic of the (Robertson–Walker) metric

$$0 = \Delta t^2 - R^2(t)\Delta r^2/c^2(1 - kr^2).$$

From this we obtain

$$\int_{t_e}^{t_0} dt/R(t) = 1/c \int_0^{r_e} dr/\sqrt{1 - kr^2}$$

and

$$\int_{t_e + \Delta t_e}^{t_0 + \Delta t_0} dt/R(t) = 1/c \int_0^{r_e} dr/\sqrt{1 - kr^2}.$$

Because the right sides of the two equations are equal, so too are the left sides. Therefore,

$$\int_{t_e}^{t_0} dt/R(t) = \int_{t_e + \Delta t_e}^{t_0 + \Delta t_0} dt/R(t).$$

But for any frequency of electromagnetic radiation, the intervals between crests, Δt_e and Δt_0, are fractions of a second, during which the relative distance the galaxy has moved is negligible. Therefore,

$$\Delta t_0/\Delta t_e = R(t_0)/R(t_e).$$

So the redshift (the fractional change in wavelength) is

$$z = \lambda_0/\lambda_e - 1 = R(t_0)/R(t_e) - 1.$$

Thus we have the redshift z in terms of the relative change in the scale factor. For example, when the redshift of a distant galaxy is measured, we know that the universe was smaller then according to

$$R = R_0/(z + 1).$$

The redshift is often used by cosmologists in place of time because its relationship to the cosmic scale factors at two different epochs has the above direct relationship, whereas the actual time difference between the two scale factors depends on the cosmological constants, which were at one time highly uncertain. Besides, the cosmological redshift is directly measurable.

6 Galaxy Clusters, Galaxies, and Stars

The oceans of the world are made up of water molecules. Each has two hydrogen atoms and one oxygen atom bound together. All of that hydrogen condensed during the first few milliseconds in the terrible furnace that *was* the universe. Several hundred million years passed before the first oxygen was forged in the gentler fires of stars.

— nkg

6.1 Structure Formation

The formation of structures in the universe began at a very early time. Quarks, those most unusual of the fundamental particles, combined to form the first neutrons and protons at one 100th of a microsecond ($t = 10^{-8}$ seconds). And by the end of ten minutes, all the neutrons that had not already decayed had combined with protons to form the cosmic abundance of helium-4 and a few other light *nuclei*. Photons still interacted strongly with all of these charged particles, traveling a very small distance before being deflected. All matter and light moved together with the flow of the cosmic expansion.

After a very long time in comparison with those early events, 300 000 years later, the universe cooled to a temperature of 3000 degrees Kelvin. Electrons could combine with the free protons and those light nuclei that had formed in the first few minutes, so that free charges vanished to form charge-neutral *atoms*. Consequently, radiation and matter became *decoupled* in the sense that the primeval photons no longer interacted with individual material particles. At that moment, light began to travel freely without hindrance from matter. It remains today, just as it was then, except for the Doppler shift of each photon to lower energy caused by the universal expansion. This light (the cosmic background radiation), now mostly in the form of microwaves[1] together with the abundance of light elements and the

---

[1] As in ovens of the same name.

161

observed expansion of the universe, provides the strongest evidence of the Big Bang scenario to this point of our account.

How, from the almost featureless cloud of hydrogen and helium gas that filled the universe at this early time, did the universe we see today come to be, a universe filled with galaxies, with clusters of galaxies, all filled with stars, and eventually heavy elements and planets? First the galaxies, stars, and clusters of galaxies. . . .

Although the individual atoms of matter and photons of radiation ceased to interact electromagnetically at the time of 300 000 years, they always interact through the gravitational force with matter just as light is deflected by the mass of our Sun when it passes close by.[2] So radiation remained an important factor in the formation of galaxies out of matter up to about the time when the mass density of radiation fell below that of matter and remained forever so. This occurred when the universe had cooled to about 2000 K at a time of about a million years (page 158). In fact a million years marks the transition from an era when galaxies were formed mainly by the action of gravity on radiation, matter being dragged along for the ride — so to speak — to the matter-dominated era when the primeval radiation faded in importance. The early interaction between matter and radiation at the time when galaxies were formed mainly through the intermediary of radiation remains imprinted on the cosmic background radiation today.

Gravity draws the mass equivalent of all types of energy together, whether it be matter, its energy of motion, radiation, or whatever. But if matter and all forms of energy are uniformly distributed, especially in an infinite universe, the pull on any piece is counterbalanced by that of all the rest. This was recognized already by Newton as a problem in explaining the stars in the heavens. How then did galaxies and stars form?

Even if the gas of elementary particles that pervaded the very early universe were uniformly spread to begin with, chance fluctuations in their random motion would have created some degree of spatial inhomogeneity. And as soon as any slight inhomogeneity forms, gravity will tend to make it grow in mass by attracting less dense surrounding matter. However, the estimated time for the growth of statistical fluctuations to galactic dimensions

---

[2]The deflection of light by gravity was predicted by both Newton and Einstein. Einstein's predicted deflection of light rays coming from distant stars that pass close to the Sun is twice that predicted by Newtonian gravity. This calculation was performed by Einstein himself. The great cosmologist Sir Arthur Eddington prepared at once to measure it at the next solar eclipse, which took place in 1919. It was an early and successful test of Einstein's gravitational theory.

has been estimated to take much longer than the actual age of the universe. So there must have existed small departures from a smooth universe in the first instants of time. How they may have come about will be a later subject related to cosmic inflation.

We assume therefore, because the universe *now* is not homogeneous on the scale of intergalactic distances, that there were small inhomogeneities, or clumps, at early times. Their size would have to be very small compared to the size of the cosmic horizon at that time. Otherwise, they would disturb the evolutionary picture of the universe that we have painted in earlier chapters. Yet they must have been large enough and sufficiently diverse in scale that within a third of the lifetime of the universe or less, gravity would have had time to do its work, forming galaxies of various sizes and clusters and stars within galaxies. That galaxies did form within such a time span places constraints on the size of the early clumpiness and on the particular era, or time frame, when large regions of the gases of the universe, about 75% hydrogen and 25% helium, began to collapse around any slight clumpiness under the force of gravity to form protogalaxies.

Sir James Jeans was the first to tackle the problem of how slight inhomogeneities in the early universe grew to form the structures that we see

Fig. 6.1. Sir James Jeans, who in 1902 pioneered the theory of the formation of galaxies out of the almost featureless clouds of matter and radiation in the early universe.

now. At that time, 1902, he was able to indicate only the outlines of the solution, and it remains today an activity, largely one employing computers that are used to simulate the growth of structure from early seeds of various origins.

Clouds in the otherwise smooth primordial gas of hydrogen and helium together with a little dust produced on the surfaces of very early supergiant stars will attract surrounding matter and at first will grow. At a critical size and mass, depending on the temperature and density of the universe at that particular time, the cloud will cease to aggregate more mass, and instead begin to collapse. However, as the cloud is compressed by gravity, gas pressure builds up and resists further collapse; the higher the temperature, the greater the pressure. As a cloud starts to collapse and its internal pressure rises, it may bounce, re-expand, collapse, bounce, and so on. Or it may continue to collapse. How big must a cloud be at a given temperature in the history of the universe before it will collapse? The size must depend on the temperature because the higher the temperature of a gas, the higher the internal pressure that resists collapse.

A cloud that starts to collapse but because of its rising internal pressure halts, re-expands, starts to collapse again, and so on, is oscillating like a sound wave, which is a density oscillation in the air. A sound wave has a velocity that is finite, so it takes a certain time for the oscillation to complete one cycle and begin another. If it takes longer for the cloud to oscillate through a cycle than to collapse, then the cloud will in fact collapse. Otherwise it will oscillate until, like sound, the oscillations are damped by viscosity and become fainter.[3] Meanwhile, conditions in the universe are changing. They may change such that the time to complete an oscillation becomes longer than the time for collapse; under the altered conditions, that cloud of given size and density may then collapse.

The photons of radiation are another important ingredient in structure formation, especially at early times when radiation density exceeds mass density. As we know by now, the photons, while having no mass, nevertheless exert a gravitational attraction and are gravitationally attracted according to their *equivalent* mass given by Einstein's famous relation, $m = E/c^2 = aT^4/c^2$. So, radiation also affects the collapse of gas clouds, or the failure to collapse. Moreover, we have learned that the density of mass — the mass per cubic centimeter — of matter and radiation change differently as the universe expands. The number of particles per cubic centimeter, and

---

[3]As a cloud collapses, atoms are ionized by the increasing temperature so that some of the energy in the oscillations is transferred. This is the nature of viscosity.

therefore the *matter density*, decreases inversely as the volume increases, which we can write in terms of the scale R of the universe as $\rho_m \sim 1/R^3$. The number density of photons behaves also as that of particles, but at the same time the energy of each photon decreases inversely to the cosmic scale because of the Doppler effect. So the equivalent mass density of radiation (photons) decreases as $\rho_r \sim 1/R^4$. Because the mass density of radiation is diluted by the universal expansion at a greater rate than the density of matter, radiation is the main factor in galaxy formation at early times and becomes less important as time goes on, eventually becoming negligible.

Let us study the meaning of Figure 6.2. Consider a typical galaxy mass of 100 billion suns (like our own). Very early in its history, until about 1/10 of a year as measured from the beginning, such a galaxy can collapse. Its mass lies above the Jeans mass. At a later time, its mass lies below the Jeans mass and it remains below for many years. A cloud of that mass will oscillate until about 1 million years. After that, it can collapse because once again its mass lies above the Jeans mass. However, fragmentation into individual stars seems not possible until much later. Consider a globular cluster of 10^6 stars and about that mass in solar masses. Between 10^{-4} years and 10^6 years it can only oscillate. Prior to or after that it can collapse. However, it can fragment into stars either earlier than about 10^{-9} years or after 10^{10} (10 billion) years as compared to a universe age of only about 15 billion years. So it would seem that the globular clusters that are seen around the bulge of the Milky Way formed very early, earlier than a fraction of a second when the temperature was about 4×10^{10} K when the time was 1/16 seconds.

6.2   Cloud Collapse in the Radiation Era

In the radiation era, the density on which gravity acts to form a galaxy belongs almost all to radiation. And the Stefan–Boltzmann law that was derived at the turn of the last century tells us how to find the radiation energy density (aT^4) and hence its mass equivalent (aT^4/c^2) at any particular temperature. Of course it is the matter whose collapse may form a galaxy, but matter moves together with the radiation in the early universe, and the density of matter is negligible compared to that of radiation in the fiery beginning. By using scaling relations we can find the material mass that is embedded in radiation and hence the mass of early galaxies (Box 23). The critical mass is called the Jeans mass — the *least* mass in any era that could successfully collapse rather than oscillate. The critical density grew

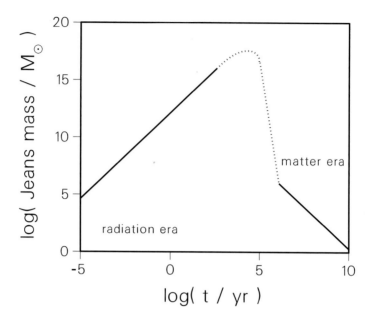

Fig. 6.2. The logarithm of the Jeans mass (minimum mass for successful collapse) measured in units of the Sun's mass is plotted as a function of time. The steep rise in the radiation era means that aside from the first year or two, only clouds of ever greater mass could collapse to form a galaxy. The situation changed radically in the matter era: ever-smaller objects could collapse to form stars.

steadily from starlike to super-galaxy-like as the universe cooled during the radiation era (Figure 6.2).

In the early radiation era, even small clouds of matter could collapse because of their high radiation content. As such a cloud collapses, its temperature rises and the very short wavelength photons can escape, thus cooling and lowering the pressure of the cloud and permitting further collapse. Thus, at very early times, before galaxies, very massive stars are believed to have formed. There is evidence for this. A distant galaxy (J1148+5251) is seen as it was only 870 million years after the Big Bang. Carbon and oxygen atoms were found in the interstellar gas, which could only have been made by still earlier stars because original atoms formed in the universe within the first few minutes of the Big Bang were only hydrogen and helium together with trace amounts of lithium and several other light elements. Carbon and oxygen — the atoms making up carbon monoxide — had to be made in the thermonuclear furnaces at the cores of the earliest stars.

However, as radiation became more dilute in the later part of the radiation era, only larger and larger clouds could collapse to form protogalaxies. The minimum mass that can collapse increases to a value of about 10^{17} times the Sun's mass in comparison with a typical galaxy mass of about 10^{11} Suns. (By way of comparison, the Milky Way mass in *visible* stars is $10^{11} M_\odot$ but there is another factor of 10 in dark matter, as can be estimated from the motion of stars in the periphery.) Larger clouds could not cool effectively; just as it takes millions of years for photons to escape from the center of the Sun, photons became trapped in the very large clouds, and they tended to fragment rather than continue their collapse to form galaxies and then stars. Galaxy formation in the late radiation era essentially came to a halt.

6.3   Matter Era

In the matter era, beginning at a million years, not only was radiation decoupled from mass, but radiation energy had been diluted by the universal expansion through the Doppler shift to such an extent that it was no longer of consequence in galaxy formation. The expansion proceeded differently with time in the two eras; whereas the cosmic gas behaves as a relativistic gas (photons) in the radiation era, it behaves as a nonrelativistic gas of hydrogen in the matter-dominated era (Box 24). In this way we learn that the minimum mass for galaxy formation in the matter-dominated era in which we ourselves live, *decreases* with falling temperature of the universe, so that smaller objects like globular clusters could form and also individual stars, whose formation has continued into our own time. These trends are depicted in Figure 6.2.

6.4   Galaxy Formation

The Big Bang ignited the universe 15 billion years ago. From the great clouds of gas of the very early universe, what came first? Did individual stars condense and gather into small galaxies, which later merged to form larger ones such as our own? Or did galaxies, as we know them now, appear first as dense clouds inside of which stars condensed? Not even the most powerful telescopes with the largest lenses that astronomers can build will ever provide an answer. But there are larger lenses — nature's *gravitational* lenses. Such a possibility was foreshadowed by Einstein's realization that mass can bend light much as an optical lens bends light to form an image. Thus it is that groups of stars or galaxy clusters can act as lenses, bending

Fig. 6.3. In this single picture, myriad galaxies can be seen, either as disks or evolving toward that shape, under the combined influence of gravity and angular momentum conservation, just as Laplace. divined (see page 3), though he lived centuries too early to see through the eyes of the Hubble Space Telescope.

the light of stars and galaxies at much greater distances than can be reached with human telescopes.

Abell 2218 is the name of a cluster of hundreds of galaxies; their combined gravity is so strong that it magnifies the light of galaxies that are located far behind it. Almost all of the galaxies in the picture, Figure 6.4, belong to this cluster, but the arcs that can be seen do not: the arcs are the distorted images of a very distant galaxy population extending 5–10 times farther than the lensing cluster (50 times fainter than objects that can be seen with ground-based telescopes). This population existed when the universe was just one quarter of its present age. Sometimes, instead of arcs, the gravitational lens makes a double image of a *single* far-off galaxy, as shown in the inset. There is another striking feature about the pregalactic object shown in the inset: the home of the first stars is tiny. It contains roughly a million stars as compared with the 400 billion stars in our galaxy. Today's mature galaxies appear to have grown from many swarms of stars like this.[4]

------

[4]The origin of the first galaxies and stars is still a matter of debate, and contrary views are held by some observers; cf. R. Barkanan and A. Loeb, *Nature*, Vol. 421 (2003), p. 341.

Fig. 6.4. In this one picture more than ten billion years of cosmic history is recorded. Modern galaxies are in the foreground (closer means later in time from the beginning). In the magnified image in the inset to the right, an object at a distance of 13.4 *billion light years* is seen distorted into two nearly identical, very red "images" by the imperfect gravitational lens created by the foreground galaxies. Were it not for the gravitational lensing of the foreground galaxies, which constitute a rich cluster called Abell 2218, the distant galaxies would be invisible to any man-made telescope. The magnified object contains only one million stars, far fewer than a mature galaxy. Scientists believe it is a very young star-forming system. Small galaxies such as this (of low mass at early cosmic times) are likely to be the objects from which present day galaxies have formed. *Courtesy of ESA, NASA, Richard Ellis (Caltech, USA) and Jean-Paul Kneib (Observatoire Midi Pyrennes, France).*

The gravitational lens allows us to see back in time to what is likely to be an early generation of stars — stars that formed several hundred million years after the emergence of the universe.[5] These early stars were true giants, with masses of 100–1000 solar masses — they are unlike any stars that exist today. Their surface temperatures would have been 20 times hotter than our Sun. They could live only a very short time — several million years for the lighter and only 10 000 for the heavier. Gravity, acting on their great mass, would have crushed them quickly, triggering a sequence of thermonuclear fusion reactions such as take place in more ordinary stars, but faster. Then they would have exploded, leaving behind a solar mass

---

[5]The object in the inset of Figure 6.4 contains only very young stars, as evidenced by the absence of the elements that stars of later generations would have produced.

neutron star or a somewhat heavier black hole, and expelling the rest of the star into the universe. Very important also, these giant stars created on their surfaces dust particles, agglomerations of molecules, which were loosed into the universe during solar flare-ups. At a much later epoch these particles became condensation points in collapsing molecular clouds around which stars like our Sun formed.

How long does it take for a galaxy to form when a great diffuse cloud begins to collapse to form a galaxy? The timescale for collapse is $t \sim 1/\sqrt{G\rho}$ (Box 22). The mean matter density *within* galaxies themselves is approximately

$$\rho_{\text{galaxy}} = 10^{-24} \text{ g/cm}^3.$$

But this is the final, not the initial density, as needed in the formula. Nevertheless, let us proceed. We find therefore that the time it takes for a dilute cloud of that density to collapse is about 100 million years.[6] How much less was the actual cloud density before it collapsed? The present average (baryon) density is 10^{-31} g/cm^3, i.e. at a time of about 15 billion years. At a time of 1 billion years, the density would have been $(15/1)^3 \sim 10^3$ larger. Presumably the density of clouds would not have been larger. If this is so, then the collapse time would be $15^{3/2} \sim 60$ shorter — in other words, galaxy formation time is less than 100/60 million years, or roughly 20 million years. Later galaxy formation would have taken longer, perhaps 100 million years for the Milky Way.

6.5   Galaxy Types

> All the effects of Nature are only the mathematical consequences of a small number of immutable laws.
>
> — Simon-Pierre Laplace

Island universes — this is what William Herschel called them when first he realized that the faint luminosities that he perceived in his telescope were enormous collections of stars that seemed small only because they were so far away. Now we know that there are up to 100 billion stars in quite ordinary galaxies, and 10 times that in some. They take several forms. Our own galaxy, the Milky Way, is a spiral arm galaxy, much like Andromeda, (Figure 1.12), and like the edge-on view of the beautiful giant warped spiral

---

[6]Newton's constant has the value $G = 6.7 \times 10^{-8} \ 1/(\text{s}^2 \cdot \text{g/cm}^3)$.

galaxy in Figure 1.3. The central bulge — a spheroid of old stars — is a prominent feature of spirals and is important for their stability. Without it, the spiral arms would spiral away. The spiral arms of NGC3370 are clearly visible in Figure 6.5. Several more distant galaxies are also visible in the figure, including an edge-on view of a spiral and an elliptical.

Fig. 6.5. A spiral arm galaxy called NGC3370, which is similar to our Milky Way galaxy, which we can see only from the inside. In the same view other more distant galaxies are plainly visible. *Credit: NASA, The Hubble Heritage Team, and A. Riess (STScI).*

The Milky Way has a mass of about 100 billion solar masses in visible stars and is a member of a cluster of some 2000 galaxies called the Virgo cluster. And it seems to be in the process of gathering to itself two small nearby galaxies, the Large and Small Magellanic Clouds. They are visible to the naked eye in the southern hemisphere as very large (compared to constellations) faint luminosities. Besides these two, there are numerous dwarf galaxies that are also bound to our own. Dust is common in most galaxies, certainly in our own, where it obscures the other side of the galaxy, but from some the gas has been stripped by a near-collision with another galaxy.

Elliptical galaxies are often much more massive than our own spiral galaxy — sometimes by a factor of 10. They range in shape from spheres to spheroids with a 3:1 ratio of the axis. A spherical elliptical is prominent in Figure 6.6. The distribution of stars is dense in the center and gradually fades toward the edge.

Fig. 6.6. An unusual view showing a bright elliptical galaxy, NGC 4881 (*upper left*), in the Coma cluster and a number of distant galaxies of several types, spirals and ellipticals, including a possible merger (*far right, just above center*). *Credit: NASA.*

6.6   Star Formation

Stars are incubated in huge gas clouds inside galaxies, sometimes singly, but often in batches. The clouds consist mostly of hydrogen and helium but with dust particles that originated on the cooler surface of red giants, those stars that are approaching their end before the explosion of their cores. The clouds are diffuse and highly nonuniform, with clumps and filaments dispersed throughout them. Their masses range between 10 and 10^7 suns and a compression of the order of 10^{20} is involved in forming a star from the diffuse gas. Particularly active star formation regions can usually be identified by their bluish color, indicating tremendous heat. The birthing of stars within our own galaxy is visible in Figure 6.7.

6.7   Nova and Supernova

The lightest elements — deuterium, helium, and lithium — were created during the first five minutes in the intensely hot universe following the Big Bang. Never since that fiery beginning has the universe been so hot. All the other elements were created at much lower temperatures and starting at several hundred million years later in stars. In fact, astrophysicists know

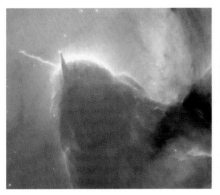 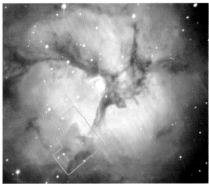

Fig. 6.7. Enlargement of a stellar nursery (left panel) of the upper right rectangle of the region outlined in blue in the Trifid Nebula (right panel). The nursery is being uncovered by radiation from a nearby, massive star about 8 light years away (central bright spot in right panel). The Trifid is about 9000 light years from the Earth, in the constellation Sagittarius. A stellar jet (the thin, wispy object pointing to the upper left) protrudes from the head of a dense cloud and extends three quarters of a light year. The jet's source is a very young stellar object that lies buried within the cloud. Jets such as this are the exhaust gases of star formation. The "stalk" (the finger-like object) points from the head of the dense cloud directly toward the central star (right panel) that illuminates the Trifid Nebula. This stalk is a prominent example of an evaporating gaseous globule (EGG). It has survived because at its tip there is a knot of gas that is dense enough to resist being eaten away by the powerful radiation from the central star. *Credit: (1) NASA and Jeff Hester (Arizona State University); (2) ground-based image of the whole nebula from Palomar, of which (1) is a very small portion of the green boxed area.*

in considerable detail how the heavier elements are forged in the interiors of massive stars during their lifetime, and how the supernova explosions at the end of a 10-million-year lifetime spew forth the small store of these elements and create the heaviest ones beyond iron in the expanding inferno outside the nascent neutron star that forms at the center. This cycle has been repeated time and again as new stars are born and old ones die, each generation enriching the universe with more of the heavy elements from which planets and eventually life were formed.

The light and heat that stars radiate are created by thermonuclear fusion in their interiors. Normally nuclei repel each other because of the positive charge on their protons. However, because of the large mass of stars, gravity compresses the gas from which they are made. The high pressure and density toward the center surmount the repulsion and fuse helium into heavier elements, like carbon and oxygen, and, in very massive stars, all the way to iron. Together, the masses of the three helium nuclei that are fused to form carbon have more mass than the carbon they make. The excess mass is converted into heat energy. Such reactions are the source of the radiant heat energy of stars, and the pressure due to the heat is what prevents gravity from immediately collapsing the star. Some of the heat does reach the surface and is radiated into the universe, producing the light and warmth we receive from the Sun, and the light we see at night from distant stars. As a result of this loss, the star sinks slowly toward its final great explosion, called a nova or a supernova. In the fiery debris of elements that is cast off, some of the iron and nickel is further processed to make the trace amounts of still heavier elements.

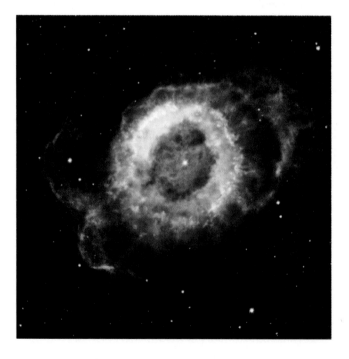

Fig. 6.8. Little Ghost (NGC 6369), a dying star, which was similar to our Sun, is expanding into the red giant stage. The star is being reborn as a white dwarf (dot at center). In the process the great cloud of gasses that now surrounds it is escaping into space and is called a planetary nebula (an historical term with no relation to planets). The gases are escaping at 15 miles per *second. Credit: Hubble Heritage Team (STScI/AURA/NASA).*

6.7.1   *Nova: creation of a white dwarf*

Stars like our Sun, which in the pantheon of stars is a small star, live to a great age of ten billion years or so. Their weak gravity acts slowly to crush the star, burning hydrogen into helium, and helium into carbon and oxygen. In old age, which amounts to billions of years, they begin to belch, so to speak. They puff up under the heat pressure in their cores to become a red giant. Through its expanded surface the star more readily loses heat and cools. As a result the pressure that caused it to expand in the first place is weakened, and it begins to contract again. Thereupon, the core gets hotter, pressure rebuilds and the star re-expands. Each time it does so, some of the outermost shells of gas escape into space, thus distributing some carbon, oxygen, and magnesium into the world. Eventually, with sufficient loss of mass, gravitational attraction fails to halt the expansion; most of the outer part of the star becomes detached and expands into what is called a planetary nebula (Figure 6.8). What remains of the star settles into what is known to us as a white dwarf — a star that is very hot, and therefore appears white.

Fig. 6.9. S. Chandrasekhar as a young man at the University of Chicago. He predicted that dead stars like neutron stars and white dwarfs could be no more massive than about 1.4 times the mass of our Sun. *Credit: University of Chicago Press.*

S. Chandrasekhar (Figure 6.9), while still a youth on his way by steamer to England in 1930, there to study under Sir Arthur Eddington, formulated a theory of the strange and until-then-not-understood white dwarfs, which had only recently been discovered. He predicted that these dead stars could not have a mass greater than about 1.4 times the Sun's mass. This limit is now known as the Chandrasekhar mass limit. Eddington never believed this theory, thinking it quite absurd and saying so. This pained the young man deeply, and he turned his attention to other aspects of astrophysics, returning to the implication of his youthful theory much later in life. The implication of the theory was that a dead star of greater mass than the one he calculated would begin to sink forever into a black hole, a fate that no one had as yet foreseen. Eddington apparently did catch the implication but did not accept it as a possible natural outcome.

We now know that Chandrasekhar was correct in his prediction of a maximum possible mass for dead stars (including of course, neutron stars). And Eddington was correct in his conclusion, which he nonetheless rejected as unphysical, that if such a limit were to exist, it would imply that more massive dead stars would collapse to form black holes.

6.7.2   *Supernova: creation of a neutron star or black hole*

In the interior of massive stars, ten times or more the mass of our Sun, the density is so great that atoms are crushed to the point where their electrons are stripped away and they move frantically in the interstice between the nuclei, thus creating a high pressure. The nuclei themselves encounter each other at high speed and overcome the resistance of the Coulomb force attributable to their electric charge. In this way nuclei fuse in what are called thermonuclear reactions. The fusion of light elements in the original star — hydrogen and helium — produces heavier nuclei — carbon, oxygen, all the way up to iron. And the fusion of the lighter elements produces some heat energy and pressure, which act to resist the collapse of the star. But at each stage of fusion, the weight of the star compresses the iron core. Fusion ceases at iron because that is the heaviest nucleus whose fusion produces an output of heat energy, the very energy that sustains the star against gravitational collapse. The elements — iron, oxygen, and carbon — sink to the center and form concentric shells, the heaviest at the center.

As more of these elements are fused and sink, the mass of the *inert* iron core grows to the critical Chandrasekhar mass limit. In that instant, the internal pressure exerted by the electron motion in the compressed iron core is no longer able to sustain the core against its own gravity. In a matter of seconds, it collapses from a radius of a few hundred kilometers to about ten kilometers. At that point nucleons are brought into such proximity that they resist further compression.

Thus it is that electron pressure sustains white dwarfs at the size of hundreds of kilometers, and nucleon pressure sustains neutron stars at a radius of about ten kilometers, and in both cases at a maximum mass of about the Chandrasekhar limit of 1.4 suns. The limit for neutron stars is several tenths of a solar mass larger because the repulsion between nucleons at a very close distance adds to the resistance against collapse.

But, if the some of the material that initially was expelled in the explosion that cast off most of the star from the nascent neutron star falls back so that its mass exceeds the Chandrasekhar limit, the neutron star will begin an eternal collapse into a black hole from which nothing can ever return.

Fig. 6.10. The black hole in the center of the elliptical galaxy M87 ejects a jet of electrons and other subatomic particles traveling at nearly the speed of light. They create a blue track which contrasts with the yellow glow from the combined light of billions of unseen stars and the faint, yellow, pointlike globular clusters of stars, each containing hundreds of thousands of stars, that make up this galaxy. *Credit: NASA and The Hubble Heritage Team (STScI/AURA).*

A black hole that is tearing apart and ingesting other stars in the elliptical galaxy M87 (Figure 6.10) is firing a jet of plasma into space. The power of the radiation and its concentration in a small region of the galaxy, as well as the emission of jets of material spewed out at speeds close to light, are what reveal this shocking cannibalism. The enormous mass of the black hole, over 2.6 billion times the mass of the Sun, concentrates stars near the center of M87 with a density at least 300 times greater than expected for a normal giant elliptical, and over a *thousand* times denser than the distribution of stars in the neighborhood of our own Sun. A much smaller black hole, known as Sagittarius A, with a mass of about 3 million suns, dominates the center of our Milky Way galaxy.

6.8   Boxes 21–25

---

21   Collapse Time

We roughly estimate the collapse time of a spherical cloud of dust (no pressure) as follows. Equate gravitational potential energy of the outer shell of mass Δm with an average kinetic energy,

$$GM\,\Delta m/r = (1/2)\Delta m\,v^2\,,$$

to get

$$\tau = r/v = \sqrt{3/8\pi G\rho} \sim 1/\sqrt{G\rho}\,.$$

---

22   Jeans Mass

In an otherwise uniform universe of hydrogen and helium, imagine a slight clumpiness here and there. Focus on one of dimension R. As it begins to collapse under its own gravity and its pressure rises, it may bounce, re-expand, collapse, bounce, and so on. This oscillation has a period R/v, where v is the velocity of sound in the clump. If the period is greater than the *characteristic* time for the collapse (estimated in the previous box), the clump will collapse; otherwise it will oscillate.

Therefore, gravitational collapse can occur only if the following condition is satisfied:

$$R/v > 1/\sqrt{G\rho}\,.$$

From the value of R given by this condition, we find the volume of the cloud, and from its density we obtain a mass that is referred to as the Jeans mass:

$$M_J = 4\pi\rho v^3/3(G\rho)^{3/2}\,.$$

23 Jeans Mass in the Radiation Era

The equivalent mass density of radiation on which gravity acts is

$$\rho_r = aT^4/c^2 \,.$$

The pressure exerted by the radiation resists gravitational collapse, so we need to find under what circumstances of density and temperature gravity wins.

We can find the pressure of the radiation by noting first that 1/3 of it is moving in any particular direction with speed of light c. The momentum of this 1/3 of the radiation in a unit volume is $\rho/3 \times c$. Pressure is the momentum striking unit area per unit of time so that the radiation pressure is $(\rho/3 \times c) \times c$, or

$$p = \rho c^2/3 \,.$$

Now we can find the velocity of sound in the cloud from

$$v = \sqrt{dp/d\rho} \,.$$

These results can now be used in the expression above to find the Jeans mass in the radiation era:

$$M_J = \rho c^6/[(3T^6(3Ga)^{3/2}] \,.$$

This mass is made up mostly of radiation, which will eventually escape from the protogalaxy. So we want to know the mass of matter that it contains.

The condition that matter and radiation were equally dense in mass, $\rho_m(t_R) = \rho_r(t_E)$, defines an era $t \leq t_E$, the era of radiation dominance, during which the *matter* density can be written in terms of the *radiation* temperature using relations derived in Box 10:

$$\rho_m(t) = \rho_m(t_E) \left(\frac{R(t_E)}{R(t)}\right)^3 = \rho_m(t_E) \left(\frac{T(t)}{T(t_E)}\right)^3 = \frac{a}{c^2} T(t_E) T^3(t) \,.$$

We use this expression for the density of *matter* during the radiation era in terms of the radiation temperature at the time when the density of radiation and matter became equal. We obtain

$$M_J(t < t_E) = \frac{K}{T^3} \sim t^{3/2}, \quad \text{where } K = \frac{c^4 T(t_E)}{3a^{1/2}(3G)^{3/2}} \,.$$

24   Jeans Mass in the Matter Era

The gas that fills the universe in this era is approximately a monatomic gas of hydrogen having a specific heat ratio $\gamma = 5/3$. The pressure and sound velocity are given by

$$p = \rho_m k T_m / m_H$$

and

$$v = \sqrt{dp/d\rho_m} = \sqrt{kT_m/m_H}\,,$$

where m_H is the mass of a hydrogen atom. Using these in the general expression for the Jeans mass, we find

$$M_J = (4\pi/3)(kT_m/Gm_H)^{3/2}\rho_m^{-1/2}\,.$$

We want to express the matter temperature and density, T_m and ρ_m, in terms of the radiation temperature, T, alone. At times before and up to t_E, the matter and radiation temperatures are the same; this allows us to evaluate the constant in the condition for adiabatic expansion, $T_m V^{\gamma-1} = $ constant. We find

$$T_m(t) = T(t_E)/[\rho_m(t)/\rho_r(t_E)]^{2/3}\,.$$

Because, at all times, both in the radiation and matter eras, $T \sim 1/R$ and $\rho_m \sim 1/R^3$, we have [recall that $\rho_m(t_E) = \rho_r(t_E)$ by definition of t_E]

$$\rho_m(t) = \rho_r(t_E)[T(t)/T(t_E)]^{3/2}\,.$$

The last two results allow us to rewrite the Jeans mass in the matter-dominated era as

$$M_J(t > t_E) = CT^{3/2} \sim 1/t\,, \quad C = \frac{4\pi}{3}\left(\frac{k}{Gm_H}\right)^{3/2}\rho_m^{-1/2}(t_E)\,.$$

25   Early Matter-Dominated Universe

The formation of protogalaxies and galaxies straddles the epoch when the universal expansion changed from being dominated by radiation to being dominated by matter. In Box 10 we derived the way in which the universe expanded in the initial fireball. We turn now to the early part of the matter-dominated era. The cosmic expansion is different in the two eras for several reasons. First, the curvature constant k and cosmological constant Λ are not necessarily negligible as they were in the early fireball, although at *early* times of the matter-dominated era before matter became very diffuse, they may be ignored. Second, the mass density of radiation and of matter scale differently as the universe expands. Because of the Doppler shift of radiation, the radiation mass density scales as $1/R^4$ but matter density scales as $1/R^3$. In particular, for matter

$$\rho_m(t)/\rho_m(t_0) = R^3(t_0)/R^3(t),$$

where t_0 is some convenient reference time, such as now, and ρ_m refers to matter, not to radiation, which is now decoupled, and dilutes as $\rho_r \sim 1/t^2$ (Box 10). However, the law that relates time and scale factor is different.

The Friedmann–Lemaître equation on page 74 becomes

$$\dot{R}^2 = [8\pi G\rho(t_0)R^3(t_0)]/3R(t).$$

Integrating from $t = 0$ to t we find

$$R(t) = [6\pi G\rho(t_0)]^{1/3} R(t_0) t^{2/3}.$$

From this we have also

$$\rho_m \sim 1/R^3 \sim 1/t^2,$$

so that the *dominant* component of mass density scales with *time* in the same way in both the radiation and matter eras. For the temperature, it is different. The radiation temperature always scales as $1/R$ so that

$$T_{\text{rad}} \sim 1/t^{1/2}, \quad \text{before decoupling},$$

whereas

$$T_{\text{rad}} \sim 1/t^{2/3}, \quad \text{after decoupling}.$$

In contrast, the temperature of matter becomes ill-defined for the universe later in the matter era, because matter condenses into clouds, galaxies, stars, and planets, each with its own temperature.

7   The Future Universe

The Second Law states that disorder always increases with time. Like the argument about human progress, it indicates that there must have been a beginning. Otherwise, the universe would be in a state of complete disorder by now, and everything would be at the same temperature.

— Stephen Hawking

7.1   Dark Matter, Dark Energy

The long search for the fundamental particles of nature and the many other composite particles, from the Greek atomists through to Madame Curie, J.J. Thompson, Earnest Rutherford, Fred Reines, and many others, paved the way for Murray Gell-Mann's elucidation of the quark structure of the baryons. Every atomic nucleus is composed of the lightest two *baryons* — the proton and the neutron. These are surrounded at great distance by electrons, equal in number to the protons. The natural elements range all the way from hydrogen, with one baryon, to the heaviest, plutonium, with 244 baryons. The Earth and everything on it, the Sun and all the stars are made from these elements, and almost all their mass resides in the baryons, the electrons being negligible by comparison. We call this familiar and pervasive type of matter *baryonic matter*.

We know quite precisely the density of baryonic matter in the universe today. It amounts to five baryons per cubic *meter*.[1] We learned this by comparing the computed deuterium production in the early universe with the measured primordial abundance of elements (Section 5.4.5). But when astronomers search the heavens for stars and galaxies of stars, they find that the density of *visible* baryonic matter is much less — only about 10%. The rest we call *dark baryonic matter*. Where is it, this dark baryonic matter? We cannot be sure. Presumably it is in brown dwarfs — small stars that burn too dimly to be seen — and in distant white dwarfs and neutron stars.

---

[1]Stated otherwise, the baryon density is 3.5×10^{-31} grams per cubic centimeter.

But whatever, we know how much there is and we know we can see only a small part of it.

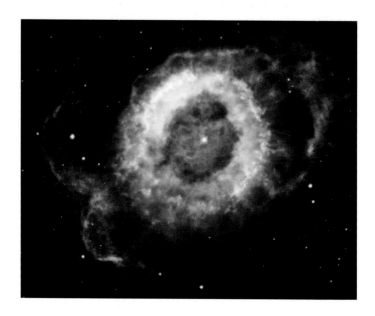

Fig. 7.1. Most white dwarfs are too dim to see. But at a certain stage in their late life they erupt into a planetary nebula (not related at all to planets). Here a white dwarf is in its red giant stage, producing the Little Ghost Nebula (NGC 6369) by ejecting most of its mass in a series of puffs. Our Sun will share this fate in about 7 billion years. The bright spot in the middle is the remnant of the original star, which is sinking into its white dwarf stage. Because of its high temperature the dwarf star will remain in view for several billion years. However, the expelled gases that we see surrounding it are traveling away at a speed of about 25 kilometers a second. The tiny white dwarf bathes the gases with X-rays. The X-rays promote electrons to higher energy states of their host atoms. We see the light (of emission lines) that is radiated by electrons when they fall to their normal state. The surrounding nebula will dissipate in about 10 000 years and the bare white dwarf will remain clearly visible. *Credit: NASA and The Hubble Heritage Team (STScI/AURA).*

However, the mystery is still deeper — it is now known that there is a great deal more matter in the universe than there is in *baryonic* matter, both visible and dark. In fact, Fritz Zwicky in the 1930s examined the relatively nearby cluster of galaxies, the Coma cluster. By means of the Doppler shift he found that the velocities of the galaxies within the clusters were a factor of 10 larger than the escape velocity (the velocity that is required to overcome the deceleration of gravity) based upon the baryonic matter actually visible in the cluster.

The particles that make up this *nonbaryonic dark matter* are quite mysterious. They do not interact either with radiation or with individual baryons, nuclei, or atoms; if they do, the interaction is so feeble that its effect on ordinary matter cannot be detected. Even the most sensitive experiments designed to detect them so far have failed. Of course, nonbaryonic dark matter does interact gravitationally with ordinary matter, as all forms of matter or energy must. Therefore, nonbaryonic dark matter must play its part in the formation and dynamics of galaxies and clusters. However, cosmologists have not yet understood its role, which would be different according to whether dark matter consists of very light and therefore relativistic particles at the time of decoupling, or heavy particles. These are referred to as hot and cold dark matter.

The problem of the mysterious particles of nonbaryonic dark matter has been joined by another, much more mysterious one — *dark energy*.

7.2   The Three Ages of the Universe

The expansion of a very hot and dense universe began about 15 billion years ago. From that beginning, the early history of the cosmos can be *inferred* from tested laws of physics. But, from the time of the synthesis of the light elements, beginning several minutes later, cosmic history can be *read* from physical evidence. Much of that history has been recounted in the preceding pages. As to the future cosmos, it seems that galaxies that we can now view through powerful telescopes will recede and vanish ever more rapidly from one another. For billions of years the force of gravity exerted by *radiation* and *matter* has slowed the expansion — but no longer.

Radiation and matter had their strongest influence in distinct eras because the density of mass behaves differently for matter than for radiation during cosmic expansion. The photons of radiation are Doppler-shifted by the expansion, so their energy decreases as $1/R$. Therefore, their equivalent mass density decreases with expansion as $\rho_r \sim 1/R^4$. But particle masses remain unchanged with time, and so their mass density varies in the same way as their number: $\rho_m \sim 1/R^3$. Because of these behaviors, radiation is diluted by the universal expansion faster than matter. This difference marks two ages, the *radiation age* and the *matter age*. Einstein's equations describe the expansion, which is controlled by deceleration in both these ages (Box 26).

However, because density is diluted by the expansion, there is a *third age*, an age that is dominated by the cosmological *constant* Λ. This constant, first proposed by Einstein and then discarded, is essentially a mystery. It is referred to as *dark energy*, with "dark" referring to the fact that it is unseen. Only its influence on the expansion of the universe is manifest. All observational evidence gathered in the last several years points to this age as the one we have recently entered.[2] Observations of distant supernovae, of minute variations in the temperature of the cosmic background radiation, as well as certain other observations of distant galaxies, have affirmed this view. The future is controlled by a pervasive and unchanging energy density that continuously fills the expanding universe.

Exponential expansion commenced several billion years ago and the rate is *accelerating*, pushed by the mysterious dark energy. We are now rushing into a new future. This was quite unexpected. As time goes on, galaxies will fade from one another's view. Our cosmic horizon will embrace fewer and fewer of the distant early galaxies that we can presently see. How was this discovered?

7.3   The Great Cosmology Experiments

The universe is not made, but is being made continuously. It is growing, perhaps indefinitely... .

— Henri Bergson,[3] *Creative Evolution* (1907)

The expansion of the universe is controlled by three *measurable* properties: the curvature parameter k, which describes the large scale curvature of space; Einstein's cosmological constant Λ, which represents the dark energy; and all types of mass density ρ — radiation, baryonic matter, both visible (as in stars and galaxies) and dark, as well as nonbaryonic dark matter.[4] The famous Hubble parameter, which gives the velocity of recession of the galaxies in terms of their distance, provides a measure of the present age of the universe. It is about

$$t_{\text{universe}} = R_0/V_0 = 1/H_0 \approx 15 \text{ billion years}.$$

---

[2]For many years, essentially since the time of Hubble's discovery of the cosmic expansion, it was believed generally that such an age as this did not really exist; that the cosmological constant was zero.

[3]Nobel Prize for Literature, 1927: "In recognition of his rich and vitalizing ideas and the brilliant skill with which they have been presented."

[4]See the Friedmann equation on page 74.

Consequently, it is woven into the value of some of the other parameters. These have been of great interest for a number of years. Many small groups or individuals have attempted to make astronomical measurements of them. However, we have had to await the advent of "big science" — involving international collaborations of many scientists and engineers with large and very fast computers — to discover their values, and hence to provide a more complete picture of the evolution of the universe and its ultimate fate. This would not have been possible without the support and technological skill of agencies such as the National Science Foundation (NSF), the National Aeronautics and Space Administration (NASA), and the Department of Energy (ERDA) in the US, as well as like agencies in Europe and Japan. These big collaborations have provided the keys to unlocking the crucial parameters that describe the nature of the universal expansion and its future.

7.3.1 *Supernova cosmology*

In 1992 the young Saul Perlmutter (Figure 7.2) was chosen to lead the Supernova Cosmology Group at the Lawrence Berkeley National Laboratory.

Fig. 7.2. Saul Perlmutter, leader of the Supernova Cosmology Group at the Lawrence Berkeley National Laboratory, who discovered that the cosmos will expand forever at an accelerating rate, pushed on by dark energy represented by Einstein's cosmological constant Λ. *With permission of S. P.*

He had previous experience in supernova searches under Rich Muller. But now Perlmutter was striking out as leader of an ambitious mission that he and his colleague Carl Pennypacker had begun a few years earlier. The intent of the mission was to penetrate into the deep past, but long after the primordial elements had been made and even after galaxies and stars had first formed, to see how the visible universe was expanding at a time when it was only about half the size it is now. Little did he, or anyone else, suspect that this venture would lead to such a strange and unexpected discovery.

Yet a sister project (the High-Z Supernova Search, led by B. Schmidt of Mt. Stromlo Observatory, Australia, and Harvard University) also provided similar evidence for *accelerated* expansion. And other types of observations have since come to the same conclusion — the expansion of the universe is speeding up, contrary to all expectations that the gravitational attraction of mass would be slowing it down. What could cause this acceleration but a uniformly distributed and constant energy density that fills the space of the universe as it expands? Without a doubt, this discovery ranks with Hubble's discovery of the expansion of the universe in the first place (page 21), and the discovery of the cosmic background radiation by Penzias and Wilson which confirmed its once-very-hot past — at a time when it was a thousand times hotter and smaller than it is now (page 67).

One of the early problems that had to be solved in cosmology was how to measure distances to very far-off objects. Sprinkled among the heavenly bodies are certain of nature's gifts to cosmologists called *standard candles*. These are objects that give off the same amount of light under all circumstances. Ordinary stars are not standard candles. Young ones are brighter than older ones of the same size. And they come in an enormous range of sizes, from 1/100 the mass of our Sun to 100 times its mass.

However, there is one type of star called a Cepheid variable, which behaves as a standard candle. This type of star varies in brightness in an oscillating fashion, because the star itself is oscillating in size in the course of days. In a complicated way that we need not know here, the distance to these stars can be determined by the relationship in the variation of color and Doppler shift. Fortunately, the nearest of these stars lie close enough that their distance from us can be measured by another means. The distance to the closest of stars can be measured by parallax, the same means by which we can judge distance by the two slightly different views provided by our two eyes. At the same time, the range in distance from us in which Cepheid variables can be seen is very large. It overlaps a part of the much larger range in which another type of event provides a standard candle.

The overlap of different methods of distance measurement provides what is referred to as a *distance ladder*. Supernovae of a special kind called *type Ia supernovae* are standard candles that extend the distance measurement over an enormous range.

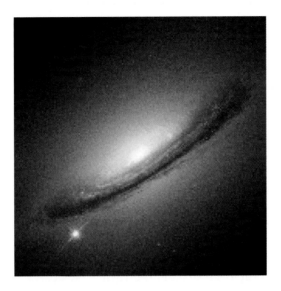

Fig. 7.3. A type Ia supernova is seen here erupting in the outer disk of a galaxy. The brightness of this explosion can be judged by its prominence in comparison with the host galaxy. Typical galaxies contain several hundred billion stars. *Credit: High-Z Supernova Search Team.*

These supernovae, the type Ia ones, are the brightest of exploding stars. They can be comparable in brightness to an entire galaxy, as seen in Figure 7.3. They can be seen to great distances and therefore to very early times in the life of the universe. They occur under special circumstances in which two stars are circling each other but of different mass — one near the Chandrasekhar mass, which is about 1.4 times our Sun's mass, and the other a lighter star. Because of the their different masses, they will burn at different rates. The heavier of the two, driven by its greater gravity, will burn faster and end as a white dwarf first. The less massive partner proceeds more slowly toward the same fate. When it reaches the stage at which it puffs up to form a red giant (see Figure 7.1), a thin stream of matter from its hot surface is attracted by the gravity of the heavier partner. Over hundreds of millions of years this slow trickle of matter drives the accreting partner toward the critical Chandrasekhar mass limit.

The partner is heated by the in-falling matter and its surface is ignited by a nuclear detonation as the critical Chandrasekhar mass is reached. It is this thermonuclear flash that is visible to great distance — billions of light years. These distant flashes (with redshifts beyond $z = 1$) occurred when the universe was less than one half of its present scale $[R/R_0 = 1/(1 + z)]$. Such flashes are called supernovae of type Ia. They are standard candles because of the uniqueness of the Chandrasekhar mass.

Perlmutter set his sights on these distant stellar explosions and made an amazing discovery. The universe, instead of being slowed in its expansion by all the matter it contains, has been accelerating for the past several billion years — in other words, *at least* since it was about 3/4 of its present scale. The simplest interpretation is that the cosmological constant Λ, which Einstein included in his early theory of gravity and then discarded, is actually real and positive. Mass and energy in the universe, if they were alone, would decelerate the universal expansion[5] because of their gravitational attraction. But a positive cosmological *constant*, which represents the dark energy, will eventually outweigh mass density when the expansion has sufficiently diluted it. Thereafter the universal expansion will accelerate, as it is doing now.

Moreover, the experiments, including two supernova projects (Supernova Cosmology, at Berkeley; and the High-Z Supernova Search, led from Australia), converge on a common result. The large scale curvature of the universe is flat,[6] curved neither like a sphere nor a hyperboloid, but exactly between (Figure 3.8). And the critical density is shared. The mass density accounts for about 30% and dark energy density for 70% (also called vacuum energy).[7] These findings are exhibited in Figure 7.4.

7.3.2   *The seed of our galaxy*

In 1989 the COBE satellite (Cosmic Background Explorer) was launched into orbit around the Earth, where its sensitive radio detectors gathered information for four years to measure the cosmic background radiation. That radiation is almost uniform from all parts of the sky but very small deviations were expected and found at the level of one part in 100 000 to a few parts per 1 000 000. These were imprinted by lumps of matter in

---

[5] See Box 8.

[6] Of course, the curvature around a black hole is extreme, but it is localized in an otherwise flat geometry.

[7] Recall that mass and radiation densities scale with expansion as $\rho_m/\rho_r \sim 1/R$ and radiation is by now negligible.

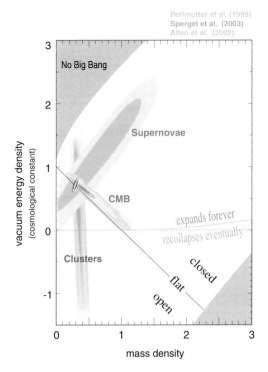

Fig. 7.4. The three great cosmology experiments alone cannot reveal the cosmic secrets. Each by itself — the supernova studies, the cosmic background radiation, and galaxy clustering, shown in different colors — is much too imprecise. However, because of the different dependence of curvature and acceleration on the cosmological parameters (Boxes 28 and 29), the regions of uncertainty intersect to yield the answer. Additionally, studies of galaxy clustering obtained from the 2dF Galaxy Redshift Survey and Sloan Digital Sky Survey share the common intersection. By these experiments the universe is discovered to have a flat spatial geometry ($k = 0$), a positive cosmological constant (dark energy) $\Omega_\Lambda = 0.7$, and matter density (of all kinds) $\Omega_M = 0.3$. *Credit: Saul Perlmutter and* Physics Today, *April 2003.*

the universe that were already present prior to the time of decoupling of radiation and matter when the universe was less than 300 000 years old (page 63).

How imprinted? As we know well by now, the universe cooled as it expanded — cooled, that is to say, on the large scale or the average — the scale on which the Einstein equations describe the evolution of the universe. But regions of slightly higher density than the surroundings — which were the seeds of galaxies — trapped radiation for a time within them.[8] Meanwhile the universe continued its expansion and the vast majority of photons continued to cool because of the cosmological redshift. So, when finally the trapped photons escaped, they were at a temperature that was higher than that of the background radiation, if only slightly (0.0003°C). The telltale signs of these early lumps remain to this very day and were detected in these exquisitely sensitive experiments. They were the seeds that eventually grew under the influence of gravity to form galaxies, clusters of galaxies, and clusters of clusters of galaxies.

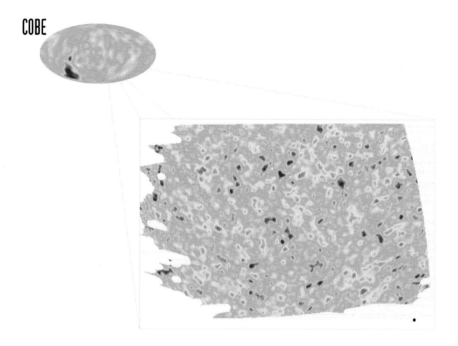

Fig. 7.5. The cosmic background radiation, of which a small portion of the whole sky, originally mapped by the COBE experiment, is here more sensitively mapped by the Boomerang experiment. Colors denote temperature, and the difference between blue (cold) and red (hot) is only 600 millionth of a degree. These small fluctuations around the present mean CBR temperature of 2.7277 K were caused at a time of 300 000 years by regions of slightly denser matter in an otherwise smooth structureless universe around which galaxies formed much later (page 144). *Courtesy: NASA and GSFC.*

---

[8]This is just as clouds in the sky trap the individual photons of sunlight, deflecting each through one droplet of water in a cloud to another until by a circuitous route — a random walk — the photon finds its way out.

Why, through these billions of years, did the temperature of these once-trapped photons remain different from the rest of the background radiation detected by Penzias and Wilson? Radiation and matter ceased to interact after about 300 000 years because all electrons became bound in atoms at about that time. As a result, there were no free charges with which photons of light could interact. Since then photons have moved through the universe as if nothing else existed (page 62).

There is an even greater wonder: conditions agreeable to life, and the presumed emergence of life from the inanimate world, occurred at a time in the history of our universe and of our civilization when cosmologists — the men and women among us who in earlier times would have been priests and priestesses — have been able to recognize the very seeds of the galaxies. One of these alone, the Milky Way, harbors billions of suns with billions of planets. From one small planet, the Earth, we look out in wonder....

7.4   Boxes 26–29

---

26   Expansion in the Three Ages

Einstein's equations describe, among other things, how gravity controls the expansion of the universe. For a uniform homogeneous universe we have seen that they take a simple form, reducing to two in number, the Friedmann–Lemaître equation and the continuity equation for energy conservation (page 74). During the radiation age, earlier than about a million years, the expansion equation takes the simple form

$$\dot{R}^2 = 8\pi G \rho_0 R_0^4 / 3R^2 \,,$$

where ρ_0 and R_0 are the values of the density and scale factor at any convenient reference time (for example the present). From this equation we learn that the size of the visible universe increases with time in proportion to the square root of time, $R \sim \sqrt{t}$, but ever more slowly. Meanwhile, the temperature falls as $T \sim 1/R \sim 1/\sqrt{t}$.

The mass densities of radiation and matter became equal at about a million years; radiation slowly faded thereafter. Equality marked the beginning of the *second age*, when the universe was dominated by matter. When ρ_m is the dominant term of those on the right of the Friedmann–Lemaître equation, the universal expansion is controlled by

$$\dot{R}^2 = 8\pi G \rho_0 R_0^3 / 3R \,.$$

In this age, the expansion increases with time as $R \sim t^{2/3}$; the speed of expansion continues to decelerate because of the gravitational attraction of matter and radiation. This is the age we have recently (several billion years ago) passed out of.

However, there is a *third age*, when density has diluted and the dark energy term Λ dominates. Let us see what its effect is; the Friedmann–Lemaître equation becomes with time

$$\dot{R}^2 = \Lambda R^2 / 3 \,.$$

The solution to the expansion equation for positive Λ, the dark energy, is $R \sim \exp \sqrt{\Lambda/3} t$. We can also note that $\ddot{R} = (\Lambda/3)R$, so that the universal expansion *accelerates* in the third era.

27   Solutions

We can summarize the time dependence of expansion as follows:

$$1 + z = \frac{R_0}{R} = \begin{cases} (t_0/t)^{1/2} & \text{radiation era}, \\ (t_0/t)^{2/3} & \text{matter era}, \\ \exp\left[\sqrt{\Lambda/3}\,(t_0 - t)\right] & \text{dark energy era}. \end{cases}$$

In the first two ages, the expansion decelerates. In the third it accelerates. The acceleration in the three ages is summarized as

$$a \sim \begin{cases} -1/(t)^{3/2} & \text{radiation era}, \\ -1/(t)^{4/3} & \text{matter era}, \\ \Lambda/3 \exp\left[\sqrt{\Lambda/3}\,t\right] & \text{dark energy era}. \end{cases}$$

28   Curvature

Recall that in cosmology we deal always with events having a common cosmic time. They lie on a surface orthogonal to all the world lines at the given time. By the cosmological principle the universe is homogeneous and isotropic. It follows that the *curvature* is everywhere the same, else one observer would see things differently than another located elsewhere in the universe.

By definition, the Hubble parameter is

$$H \equiv \dot{R}/R.$$

The Hubble equation — the first of the pair of F–L equations — can be rewritten as

$$kc^2/R^2 = H^2[\Omega_\Lambda + \Omega_M - 1],$$

where

$$\Omega_\Lambda \equiv \Lambda/3H^2, \qquad \Omega_M \equiv 8\pi G\rho/3H^2$$

are dimensionless measures of the uniform mass (or energy) density represented by the cosmological constant $\rho_\Lambda = \Lambda/8\pi G$, and of mass density of all other kinds ρ.

If cosmological measurements should find that $\Omega \equiv \Omega_\Lambda + \Omega_M = 1$, then k would have to be zero, and we would know that the universe is spatially flat. The meaning of $\Omega = 1$ can be deciphered by noting that

$$\Omega_\Lambda + \Omega_M = (8\pi G/3H^2)(\rho_\Lambda + \rho_M).$$

By defining a critical density

$$\rho_C = 3H_0^2/8\pi G = 7.7 \times 10^{-30} \text{ g/cm}^3$$

(for $1/H_0 = 15 \times 10^9$ years), we see that flatness corresponds to there being a critical density,

$$\rho_\Lambda + \rho_M \equiv \rho_C,$$

in the universe, $\rho = \rho_C$. If this density is exceeded, the universe is closed. In fact we found on page 157 that $\rho_B \approx 3.5 \times 10^{-31}$ g/cm^3. Recent cosmological evidence points to $\rho_M \approx 2 \times 10^{-30}$ g/cm^3. This suggests that there is present in the universe additional *nonbaryonic* matter of a so-far-undiscovered nature. In either case, the universe is open with an accelerating expansion.

29   Acceleration

The second of the F–L equations describes how the cosmic acceleration, $\ddot{R}$, is controlled by matter and dark energy. Until very recently (1998) it was believed that the gravitational attraction of the contents of the universe would slow the expansion. The acceleration equation can be rewritten using the earlier definitions:

$$q \equiv \ddot{R}R/\dot{R}^2 = \ddot{R}/RH^2 = \Omega_\Lambda - \Omega_M/2 - 4\pi Gp/H^2c^2 \,.$$

The last term — pressure — receives contributions only from relativistic sources like photons and neutrinos, which in the present and all recent epochs have small densities and even smaller pressure ($p = \rho c^2/3$). Matter, like baryons, and cold dark matter, even though they may dominate, are nonrelativistic and have vanishing pressure. So for the dimensionless acceleration parameter we have

$$q \geq \Omega_\Lambda - \Omega_M/2 \,.$$

Notice that the curvature constant and the acceleration parameter define intersecting lines in the Ω_Λ vs Ω_M plane. So a measurement of curvature and acceleration will determine the values of the cosmological constant *and* the nonrelativistic matter content of the universe. (Neutrinos are relativistic but not baryons.)

Books for Further Study

Nontechnical

Galileo Galilei
Sidereus Nuncius (*Starry Messenger*), translated by Albert van Helden (University of Chicago Press, 1989)

Galileo Galilei
Discoveries and Opinions, translated by Stillman Drake (Anchor Books, Random House, 1957)

Brian Greene
The Elegant Universe (Vintage Books, 1999)

Martin J. Rees
Before the Beginning: Our Universe and Others (Helix Books, 1998)

Martin J. Rees
Our Cosmic Habitat (Princeton University Press, 2003)

Abraham Pais
Subtle Is the Lord: The Science and Life of Albert Einstein (Oxford University Press, 1982)

A. Rupert Hall
Isaac Newton (Cambridge University Press, 1992)

Fang Li Zhi and Li Shu Xian
Creation of the Universe (World Scientific, 1989)

J. Silk
The Big Bang (W.H. Freeman and Co., 2001)

D. Berlinski
Newton's Gift (Touchstone, Simon and Schuster, 2000)

M. Sharratt
Galileo: Decisive Innovator (Cambridge University Press, 1994)

M. A. Finocchiarco
The Galileo Affair (University of California Press, 1989)

S. Hawking
A Brief History of Time (Bantam Books, 1998)

Technical

J.V. Narlicker
An Introduction to Cosmology (Cambridge University Press, 2002)

M.V. Berry
Principles of Cosmology and Gravitation (IOP Publishing, 1989)

A.R. Liddle and D. H. Lyth
Cosmological Inflation and Large-Scale Structure (Cambridge University Press, 2000)

Francis Graham-Smith and Andrew G. Lyne
Pulsar Astronomy (Cambridge University Press, 1998)

Richard N. Manchester and Joseph H. Taylor
Pulsars (W.H. Freeman and Co., 1977)

R.D. Blandford, A. Hewish, A.G. Lyne, and L. Mestel, Eds.
Pulsars as Physics Laboratories (Oxford Science Publications, Oxford University Press, 1992)

Ia. B. Zeldovich and I. D. Novikov
Stars and Relativity (Dover Publications, reissue, 1997)

Stuart L. Shapiro and Saul A. Teukolsky
Black Holes, White Dwarfs and Neutron Stars: The Physics of Compact Objects (John Wiley and Sons; 1st ed., 1983)

Norman K. Glendenning
Compact Stars: Nuclear Physics, Particle Physics, and General Relativity (Springer-Verlag; 1st ed., 1996; 2nd ed., 2000)

Index